Coordinate Geometry of Two Dimensions

Coordinate Geometry of Two Dimensions

HARI KISHAN
M.Sc. (Maths), Gold Medalist, Ph.D.
Senior Reader
Department of Mathematics
Kishori Raman Postgraduate College
MATHURA

ATLANTIC®
PUBLISHERS & DISTRIBUTORS

Published by
ATLANTIC®
PUBLISHERS & DISTRIBUTORS
B-2, Vishal Enclave, Opp. Rajouri Garden,
New Delhi-110027
Phones : 25413460, 25429987, 25466842

Sales Office
7/22, Ansari Road, Darya Ganj,
New Delhi-110002
Phones : 23273880, 23275880, 23280451
Fax : 91-11-23285873
web : www.atlanticbooks.com
e-mail : info@atlanticbooks.com

Printed in India
at Nice Printing Press, Delhi

Preface

The present book *Coordinate Geometry of Two Dimensions* is primarily meant for students of B.Sc./B.A. (I year) of several Indian Universities. The underlying object of the book is to provide the reader with a detailed study of the topics included in the UGC syllabus. The book exactly covers the prescribed syllabus. Nothing irrelevant has been included nor any essential matter has been excluded. Its approach is explanatory, lucid and comprehensive. The explanation of the subject matter is very systematic which would enable the students to assess and solve the problems all the better. All fundamental concepts of the included topics have been analytically explained. Sufficient number of high-graded solved examples have been given to enable the students better understand the various skills necessary in solving the problems. Practice exercises of multiple varieties have also been given which would definitely prove to be helpful to students in gaining confidence in the understanding of the subject. The figures are drawn in an accurate manner. These enable the students to grasp the subject matter in an easier way. All the answers have been checked and verified thoroughly.

Perfection in all aspects has been our aim and all efforts have been made to make the book error-free. However, if the readers point out any mistakes, which might have unknowingly crept in the text, I would rectify the same in the next edition. Constructive suggestions with a view to improve the quality of the book or to better fulfil the requirements of the students are always welcome and would be incorporated in the ensuing edition.

I am thankful to my son Manish Goyal, an author and Lecturer in Mathematics, GLAITM, Mathura, for his help in the preparation of the book.

Last but not least, I express my gratitude to Dr. K.R. Gupta, Chairman, Atlantic Publishers & Distributors, for his encouragement and also for publishing the book in so nice a form.

I wish my readers brilliant success and golden future.

198/A, Kachchi Mandi,
Kosi Kalan-281403

Hari Kishan

Contents

1

Polar Coordinates, Polar Equations of a Straight Line and a Circle

1.1 Introduction

We know that the coordinates are those quantities by which we find out the position of a point (in plane or space). For this purpose, we have been using rectangular cartesian system till now. This is a very important and popular coordinate system. In this system a point is located by finding its distances from two perpendicular lines. These lines are known as X-axis and Y-axis and the coordinates of a point are represented by (x, y). But there are various types of coordinate systems. Here we shall introduce a new coordinate system in which the coordinates of a point in a plane are its distance from a fixed point and its direction from a fixed line. The coordinates given in this way are called polar coordinates. The proper choice of a coordinate system depends on the nature of the problem at hand.

1.2 Polar Coordinates

Let O be a fixed point called *pole*. Let OX be a fixed straight line called *initial line* or *polar axis*. Consider a point P in the plane.

Let $OP = r$ and $\angle XOP = \theta$. Then, the length r is called the *radius vector* of the point P and the angle XOP the *vectorial angle*. The ordered pair (r, θ) is called the polar coordinates of the point P.

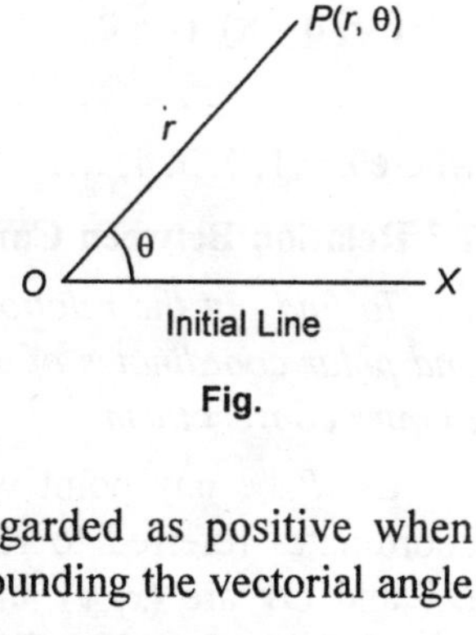

Fig.

Note 1. The radius vector r is regarded as positive when measured from the pole along the line bounding the vectorial angle and negative in the opposite direction.

Note 2. The vectorial angle θ is regarded as positive when measured anticlockwise and negative when measured clockwise from the initial line.

Note 3. If PO is produced to P' such that

$$OP' = OP$$

in magnitude and if (r, θ) be the coordinates of P, then the coordinates of P' will be either

$(-r, \theta)$ or $(r, \pi + \theta)$.

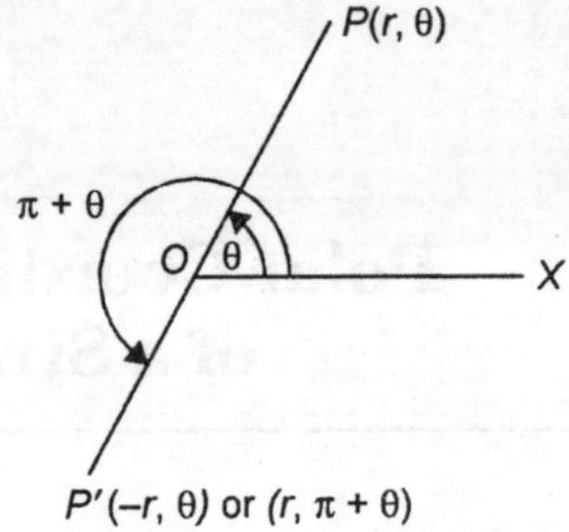

Fig.

Note 4. If the coordinates of a point are $(r, -\theta)$, it will be shown as in the adjoining figure.

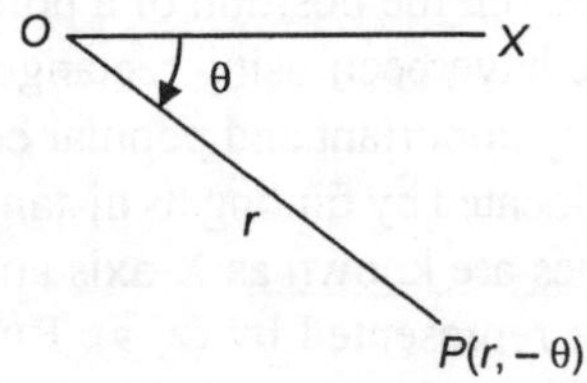

Fig.

Note 5. In polar coordinate system, unlike cartesian coordinate system, the representation of a point is not unique. In this system, a point can be specified by an infinite number of sets of coordinates. Thus if P be a point (r, θ), it can be represented as :

$(r, \theta + 2\pi), (r, \theta + 4\pi), (r, \theta + 6\pi), \ldots, (r, \theta + 2n\pi), \ldots,$

$(r, \theta - 2\pi), (r, \theta - 4\pi), (r, \theta - 6\pi), \ldots, (r, \theta - 2n\pi), \ldots,$

$(-r, \theta + \pi), (-r, \theta + 3\pi), (-r, \theta + 5\pi), \ldots, \{-r, \theta + (2n-1)\pi\}, \ldots$

$(-r, \theta - \pi), (-r, \theta - 3\pi), (-r, \theta - 5\pi), \ldots, \{-r, \theta - (2n-1)\pi\}, \ldots$

where $n = 1, 2, 3, 4, \ldots$

1.3 Relation Between Cartesian and Polar Coordinates

To find out the relations in cartesian and polar coordinates of a point in order to inter convert them.

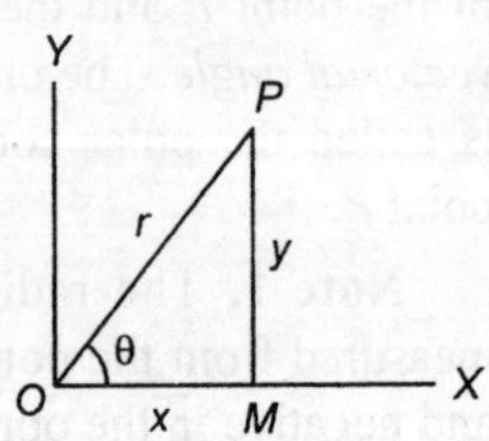

Fig.

Let P be any point whose cartesian coordinates referred to rectangular axes OX and OY are (x, y) and whose polar coordinates referred to O as pole and OX as initial line are (r, θ).

Draw $PM \perp OX$.

Then from right angled triangle OMP, we have

$$x = r\cos\theta \qquad \text{...(1)}$$

$$y = r\sin\theta \qquad \text{...(2)}$$

Squaring and adding (1) and (2), we get

$$x^2 + y^2 = r^2$$

$$\Rightarrow \qquad r = \sqrt{x^2 + y^2}. \qquad \text{...(3)}$$

Dividing (2) by (1), we get

$$\tan\theta = \frac{y}{x}$$

$$\Rightarrow \qquad \theta = \tan^{-1}\left(\frac{y}{x}\right). \qquad \text{...(4)}$$

With the help of the relations (1) and (2), we can change the cartesian coordinates of a point into polar coordinates.

With the help of the relations (3) and (4), we can change the polar coordinates of a point into cartesian coordinates.

Thus cartesian and polar coordinates of a point are inter-convertible.

Note. In order to convert a relation between x and y into a relation between r and θ, we write $r\cos\theta$ for x and $r\sin\theta$ for y.

In order to convert a relation between r and θ into a relation between x and y, we write $\sqrt{x^2 + y^2}$ for r and $\tan^{-1}(y/x)$ for θ.

1.4 Distance Between Two Points

To find the distance between two points whose polar coordinates are given.

Let the polar coordinates of the points P and Q be (r_1, θ_1) and (r_2, θ_2) respectively referred to O as pole and OX as initial line. Then

$$OP = r_1, \angle XOP = \theta_1$$

$$OQ = r_2, \angle XOQ = \theta_2$$

$$\therefore \angle POQ = \angle XOQ - \angle XOP$$

$$= \theta_2 - \theta_1$$

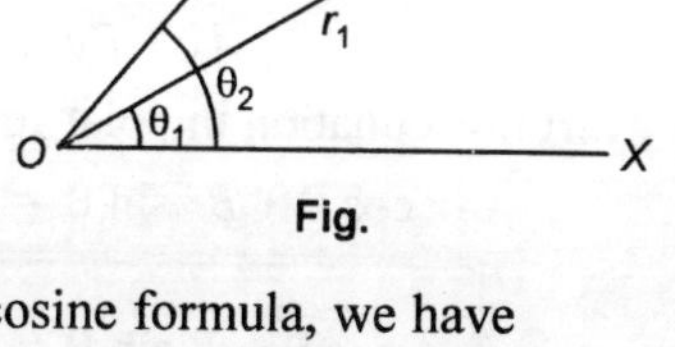

Fig.

Now from trigonometry, using cosine formula, we have

$$PQ^2 = OP^2 + OQ^2 - 2OP \cdot OQ \cdot \cos POQ$$

$$= r_1^2 + r_2^2 - 2r_1r_2 \cos(\theta_2 - \theta_1)$$

$$\Rightarrow \qquad PQ = \sqrt{r_1^2 + r_2^2 - 2r_1r_2 \cos(\theta_2 - \theta_1)}$$

which is the required distance.

1.5 General Equation of a Line

To find the general equation of a line in polar coordinates, if p is perpendicular from pole to line and α is inclination of perpendicular from initial line.

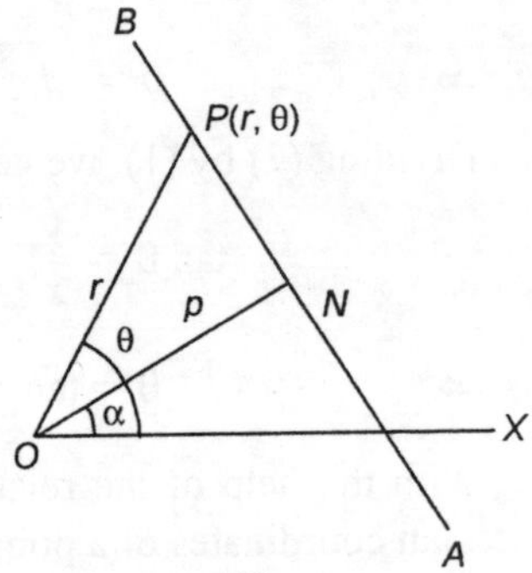

Fig.

Let O be the pole and OX be the initial line. Let AB be the straight line whose general equation is to be determined.

Draw $ON \perp AB$.

Let $ON = p$.

Let ON make an angle α with OX.

Let $P(r, \theta)$ be any point on the straight line AB.

Then,
$$OP = r$$
$$\angle XOP = \theta,$$
$$\angle XON = \alpha$$
$$\therefore \quad \angle PON = \angle XOP - \angle XON$$
$$= \theta - \alpha$$

$\therefore$ From right angled triangle ONP,

$$OP \cos \angle PON = ON$$

$$\Rightarrow \qquad \boxed{r \cos(\theta - \alpha) = p} \qquad \ldots(1)$$

which is the required equation.

Aliter : General equation of a straight line in cartesian coordinates system is,

$$Ax + By = C$$

Convert this equation in polar form, we have

$$Ar \cos\theta + Br \sin\theta = C$$

$$\Rightarrow \qquad \boxed{A \cos\theta + B \sin\theta = \frac{C}{r}}$$

which is the most general equation of a straight line in polar form.

Again, this equation can be written as

$$\frac{A}{\sqrt{A^2+B^2}}\cos\theta+\frac{B}{\sqrt{A^2+B^2}}\sin\theta = \frac{C}{\sqrt{A^2+B^2}}$$

$$\Rightarrow \qquad \cos\alpha\cos\theta+\sin\alpha\sin\theta = \frac{p}{r}$$

where $$\frac{A}{\sqrt{A^2+B^2}} = \cos\alpha, \ \frac{B}{\sqrt{A^2+B^2}} = \sin\alpha$$

and $$\frac{C}{\sqrt{A^2+B^2}} = p$$

which can be assumed because,

$$\cos^2\alpha+\sin^2\alpha = \frac{A^2}{A^2+B^2}+\frac{B^2}{A^2+B^2} = 1$$

and also, p = length of perpendicular from pole (origin) to $Ax + Bx = C$

$$= \frac{C}{\sqrt{A^2+B^2}}$$

$$\Rightarrow \qquad \cos(\theta-\alpha) = \frac{p}{r}$$

$$\Rightarrow \qquad \boxed{r\cos(\theta-\alpha) = p}$$

which is the required equation.

1.6 Different Forms of Equation of Straight Line

Corollary 1. *If a straight line passes through pole and makes an angle β with the initial line, then its polar equation is θ = β.*

This is true because the vectorial angle of any point on this line is β whereas the radius vector r may be of any measure.

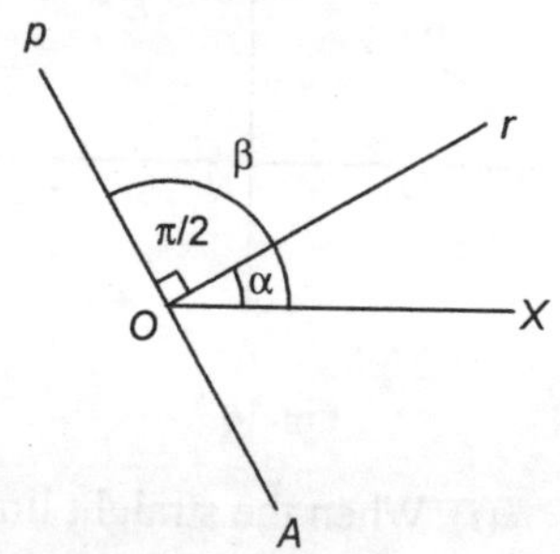

Fig.

Also, we can derive it from equation (1) as follows :

Here $p = 0$

$\therefore \qquad r\cos(\theta-\alpha) = 0$

$\Rightarrow \qquad \cos(\theta-\alpha) = 0$

as $\quad r = 0$ represents only one point i.e., pole

$$\Rightarrow \quad \cos(\theta - \alpha) = \cos \pi/2$$

$$\Rightarrow \quad \theta - \alpha = \pi/2$$

$$\Rightarrow \quad \theta = \alpha + \pi/2 = \beta \text{ (say).}$$

Corollary 2. *If a straight line is parallel to the initial line and at a distance p from the pole, then its equation is, r sin θ = p.*

Here, $\quad \alpha = \dfrac{\pi}{2}$

$$\therefore \quad r\cos(\theta - \alpha) = p$$

$$\Rightarrow \quad r\cos\left(\theta - \frac{\pi}{2}\right) = p$$

$$\Rightarrow \quad r\sin\theta = p$$

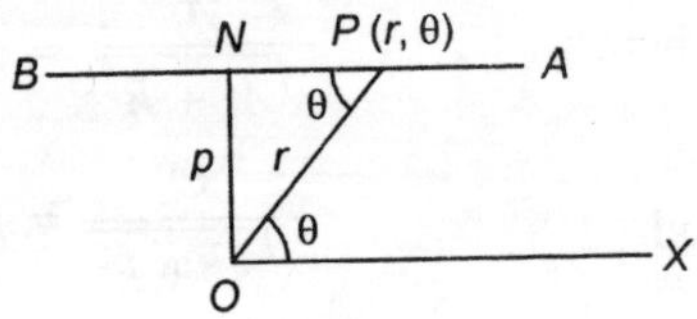

Fig.

Corollary 3. *If the straight line perpendicular to the initial line and at a distance p from the pole, then its equation is, r cos θ = p or –p according as the line is on the right or left of the pole.*

Here, $\alpha = 0$ or π according as the line is on the right or left of the pole.

$$\therefore \quad r\cos(\theta - \alpha) = p$$

$$\Rightarrow \quad r\cos(\theta - 0) = p \quad \text{or} \quad r\cos(\theta - \pi) = p$$

$$\Rightarrow \quad \boxed{r\cos\theta = p} \text{ or } \boxed{r\cos\theta = -p}$$

according as the line is on the right or left of the pole.

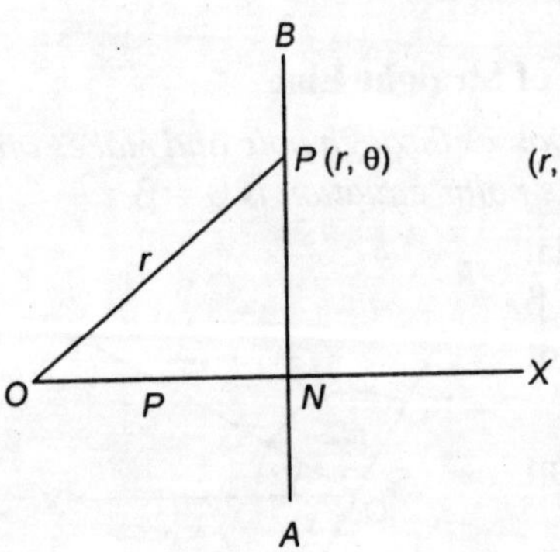

Fig. (*i*)

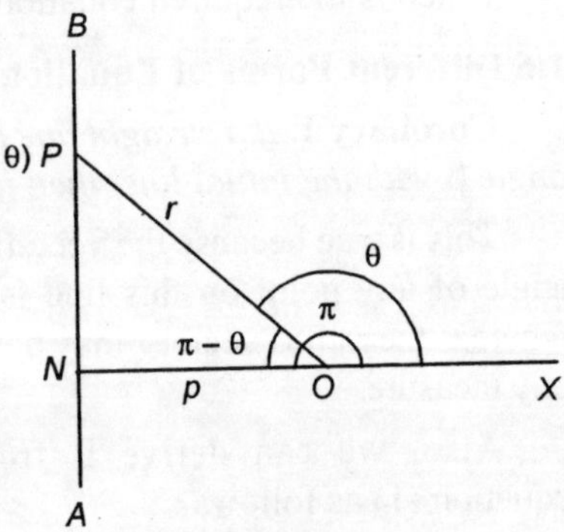

Fig. (*ii*)

(*i*) When the straight line is on the right of the pole.

(*ii*) When the straight line is on the left of the pole.

1.7 Polar Equation of a Straight Line through Two Given Points

To find the polar equation of a straight line joining two points whose polar coordinates are (r_1, θ_1) *and* (r_2, θ_2).

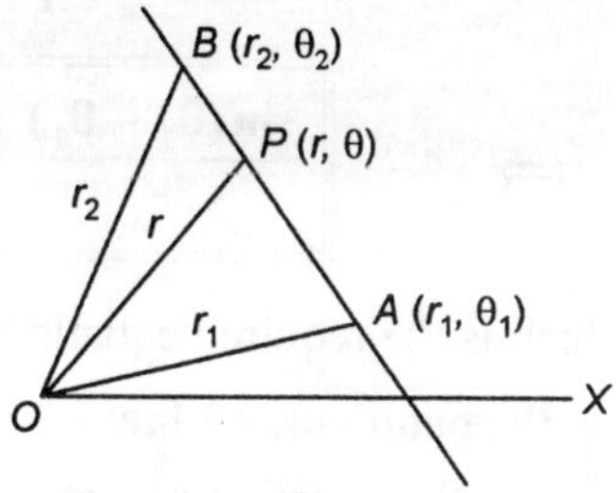

Fig.

Let the two given points be A and B whose polar coordinates referred to O as pole and OX as initial line are (r_1, θ_1) and (r_2, θ_2) respectively. Let P be any point on the line AB. Let the coordinates of P be (r, θ). Then,

Area of $\Delta\, AOB$ = Area of $\Delta\, AOP$ + Area of $\Delta\, POB$

$$\Rightarrow \quad \frac{1}{2} r_1 r_2 \sin AOB = \frac{1}{2} r_1 r \sin AOP + \frac{1}{2} r r_2 \sin POB$$

$$\Rightarrow \quad r_1 r_2 \sin(\theta_2 - \theta_1) = r_1 r \sin(\theta - \theta_1) + r r_2 \sin(\theta_2 - \theta)$$

$$\Rightarrow \quad \boxed{\frac{\sin(\theta_2 - \theta_1)}{r} = \frac{\sin(\theta - \theta_1)}{r_2} + \frac{\sin(\theta_1 - \theta)}{r_1}}$$

which is the required equation.

Aliter. Let the required equation of the line be,

$$\frac{l}{r} = A \cos\theta + B \sin\theta \qquad \text{...(1)}$$

Since it passes through $A (r_1, \theta_1)$ and $B (r_2, \theta_2)$, we have

$$\frac{l}{r_1} = A \cos\theta_1 + B \sin\theta_1 \qquad \text{...(2)}$$

and
$$\frac{l}{r_2} = A \cos\theta_2 + B \sin\theta_2 \qquad \text{...(3)}$$

Eliminating l, A and B from (1), (2) and (3) determinantically, we have

$$\begin{vmatrix} 1/r & \cos\theta & \sin\theta \\ 1/r_1 & \cos\theta_1 & \sin\theta_1 \\ 1/r_2 & \cos\theta_2 & \sin\theta_2 \end{vmatrix} = 0$$

$$\Rightarrow \quad \frac{\sin(\theta_2-\theta_1)}{r}+\frac{\sin(\theta-\theta_2)}{r_1}+\frac{\sin(\theta_1-\theta)}{r_2}=0$$

$$\Rightarrow \quad \boxed{\frac{\sin(\theta_2-\theta_1)}{r}=\frac{\sin(\theta-\theta_1)}{r_2}+\frac{\sin(\theta_2-\theta)}{r_1}}$$

which is the required equation of the line.

1.8 Perpendicular Lines

Let the equation of a straight line be,

$$\frac{l}{r}=A\cos\theta+B\sin\theta \quad \ldots(1)$$

Put $A=k\cos\alpha$ and $B=k\sin\alpha$

Then (1) yields to,

$$\frac{l}{r}=k\cos\alpha\cos\theta+k\sin\alpha\sin\theta$$

$$\Rightarrow \quad \frac{l}{r}=k\cos(\theta-\alpha) \quad \ldots(2)$$

Clearly, the perpendicular from the pole to line (2) makes an angle α with the initial line.

Let the equation of the line perpendicular to (2) be,

$$\frac{l'}{r}=k'\cos(\theta-\alpha') \quad \ldots(3)$$

Then, $\quad \alpha-\alpha'=\frac{\pi}{2}$

$\therefore \quad \alpha'=\alpha-\frac{\pi}{2}$

$\therefore$ Equation (3) becomes,

$$\frac{l'}{r}=k'\cos\left(\theta-\alpha+\frac{\pi}{2}\right)$$

$$\Rightarrow \quad \frac{l'}{rk'}=\cos\left(\theta+\frac{\pi}{2}-\alpha\right)$$

$$\Rightarrow \quad \frac{kl'}{rk'}=k\cos\left(\theta+\frac{\pi}{2}-\alpha\right)$$

$$\frac{(kl'/k')}{r}=k\left[\cos\left(\theta+\frac{\pi}{2}\right)\cos\alpha+\sin\left(\theta+\frac{\pi}{2}\right)\sin\alpha\right]$$

$$\Rightarrow \quad \frac{(kl'/k')}{r} = (k\cos\alpha)\cos\left(\theta+\frac{\pi}{2}\right)+(k\sin\alpha)\sin\left(\theta+\frac{\pi}{2}\right)$$

$$\Rightarrow \quad \frac{L}{r} = A\cos\left(\theta+\frac{\pi}{2}\right)+B\sin\left(\theta+\frac{\pi}{2}\right)$$

where $L = \dfrac{kl'}{k'}$

Hence, the equation of a line perpendicular to

$$\frac{1}{r} = A\cos\theta + B\sin\theta$$

is obtained by replacing θ by $\theta + \pi/2$ and l by a new constant.

1.9 Polar Equation of a Circle

To find the polar equation of a circle whose radius and the coordinates of centre are given.

Let O be the pole and OX be the initial line. Let $C(R, \alpha)$ be the centre of the circle. Let a be the radius of the circle.

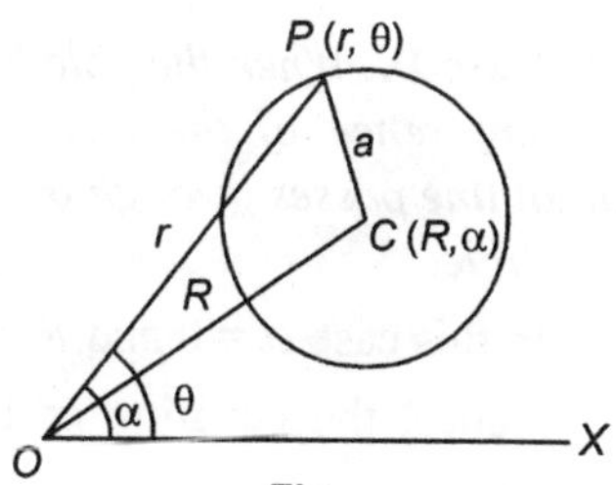

Fig.

Let $P(r, \theta)$ be any point on the circle. Join PC, OC and OP.

Then, $OC = R$, $OP = r$, $CP = a$,

$$\angle XOC = \alpha,$$
$$\angle XOP = \theta$$

$$\therefore \quad \angle COP = \angle XOP - \angle XOC = \theta - \alpha$$

Therefore, from $\Delta\, CPO$, using Cosine formula of trigonometry, we have

$$CP^2 = OC^2 + OP^2 - 2OC \cdot OP \cdot \cos(\angle COP)$$

$$\Rightarrow \quad a^2 = R^2 + r^2 - 2Rr\cos(\theta-\alpha)$$

$$\Rightarrow \quad \boxed{r^2 - 2Rr\cos(\theta-\alpha) + R^2 - a^2 = 0}$$

which is the required equation of the circle.

Particular Cases :

Case I. When centre is the pole.

In this case, $R = 0$.

Hence, the equation of the circle becomes

$$r^2 - a^2 = 0$$

$$\Rightarrow \quad \boxed{r = a}$$

Case II. *When centre lies on the initial line.*

In this case, $\alpha = 0$.

Hence, the equation of the circle becomes

$$\boxed{\boldsymbol{r^2 - 2Rr \cos \theta + R^2 - a^2 = 0}}$$

Case III. *When the pole lies on the circle.*

In this case, $R = a$.

Hence, the equation of the circle becomes

$$r^2 - 2ar \cos (\theta - \alpha) + a^2 - a^2 = 0$$

$$\Rightarrow \quad \boxed{\boldsymbol{r = 2a \cos (\theta - \alpha)}}$$

Case IV. *When the pole lies on the circumference of the circle and the initial line passes through the centre of the circle.*

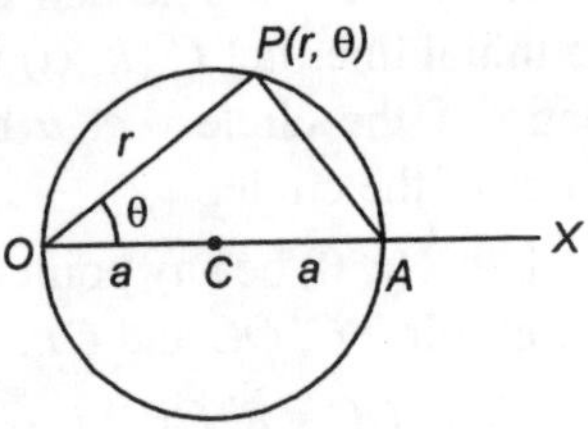

Fig.

In this case $\alpha = 0$ and $R = a$.

Hence, the equation of the circle becomes

$$r^2 - 2ar \cos \theta + a^2 - a^2 = 0$$

$$\Rightarrow \quad \boxed{\boldsymbol{r = 2a \cos \theta}}$$

Note. The general equation of the circle is,

$$r^2 - 2R\, r \cos (\theta - \alpha) + R^2 - a^2 = 0$$

This is a quadratic equation in r. For any given value of θ, it gives two values of r, say r_1 and r_2.

Let $OP = r_1$ and $OQ = r_2$

Then, product of the roots $= r_1 \cdot r_2$

$$= OP \cdot OQ$$

$$= R^2 - a^2$$

which is independent of θ.

This establishes a well-known proposition that "*If from a given point O, straight lines be drawn to cut the given circle in two points P and Q, then OP · OQ is independent of the direction of the line drawn."*

1.10 Equation of the Tangent to the Circle

To find the equation of the tangent at any point θ_1 of the circle

$$r = 2a\cos\theta$$

The equation of the circle is,

$$r = 2a\cos\theta \quad \text{...(1)}$$

Let $P(r_1, \theta_1)$ and $Q(r_2, \theta_2)$ be two points on the circle.

Then, $$r_1 = 2a\cos\theta_1 \quad \text{...(2)}$$

and $$r_2 = 2a\cos\theta_2 \quad \text{...(3)}$$

Equation of the straight line PQ is,

$$\frac{\sin(\theta_2 - \theta_1)}{r} = \frac{\sin(\theta - \theta_1)}{r_2} + \frac{\sin(\theta_2 - \theta)}{r_1}$$

$$= \frac{\sin(\theta - \theta_1)}{2a\cos\theta_2} + \frac{\sin(\theta_2 - \theta)}{2a\cos\theta_1}$$

$$= \frac{1}{2a\cos\theta_1\cos\theta_2}[\sin(\theta - \theta_1)\cos\theta_1 + \sin(\theta_2 - \theta)\cos\theta_2]$$

$$= \frac{1}{4a\cos\theta_1\cos\theta_2}[2\sin(\theta - \theta_1)\cos\theta_1 + 2\sin(\theta_2 - \theta)\cos\theta_2]$$

$$= \frac{1}{4a\cos\theta_1\cos\theta_2}[\sin\theta + \sin(\theta - \theta_1 - \theta_1) + \sin(\theta_2 - \theta + \theta_2) + \sin(\theta_2 - \theta - \theta_2)]$$

$$= \frac{1}{4a\cos\theta_1\cos\theta_2}[\sin(\theta - 2\theta_1) + \sin(2\theta_2 - \theta)]$$

$$= \frac{1}{4a\cos\theta_1\cos\theta_2}[2\sin(\theta_2 - \theta_1)\cos\{\theta - (\theta_1 + \theta_2)\}]$$

$$= \frac{\sin(\theta_2 - \theta_1)\cos\{\theta - (\theta_1 + \theta_2)\}}{2a\cos\theta_1\cos\theta_2}$$

$$\Rightarrow \frac{2a\cos\theta_1\cos\theta_2}{r} = \{\cos\{\theta - (\theta_1 + \theta_2)\}\} \quad \text{...(4)}$$

when $Q \to P$ i.e., $\theta_2 \to \theta_1$, (4) becomes

$$\frac{2a\cos^2\theta_1}{r} = \cos(\theta - 2\theta_1)$$

$$\Rightarrow \boxed{\boldsymbol{r\cos(\theta - 2\theta_1) = 2a\cos^2\theta_1}} \quad \ldots(5)$$

which is the required equation of the tangent to the circle

$$r = 2a\cos\theta$$

at the point θ_1.

Note. If the equation of the circle is $r = 2a\cos(\theta - \alpha)$, then the equation of the tangent is,

$$r\cos(\theta - 2\theta_1 + \alpha) = 2a\cos^2(\theta_1 - \alpha)$$

[Writing $\theta - \alpha$ in place of θ and $\theta_1 - \alpha$ in place of θ_1]

1.11 Area of a Triangle

To find the area of triangle if the polar coordinates of vertices of triangle are known.

Let O is pole, OX is initial line.

Let the polar coordinates of vertices A, B, C of triangle are (r_1, θ_1), (r_2, θ_2) and (r_3, θ_3) respectively.

Clearly, area of $\Delta\, ABC$ = area of ΔCOB – area of ΔCOA – area of $\Delta\, AOB$

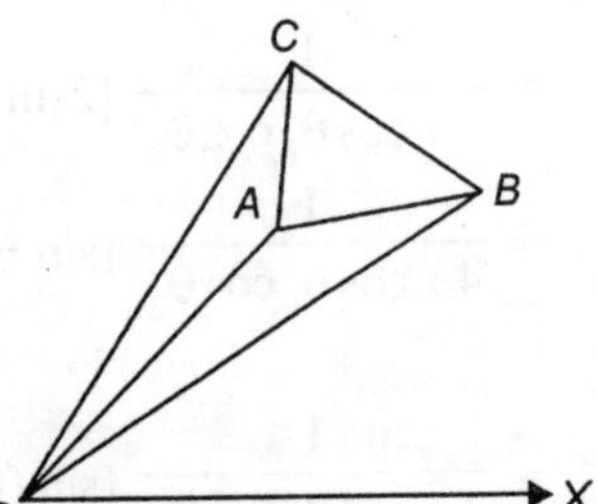

Fig.

Area of ΔCOB

$$= \frac{1}{2} OB \cdot OC \sin COB$$

$$= \frac{1}{2} r_2 r_3 \sin(\theta_3 - \theta_2)$$

Area of ΔCOA

$$= \frac{1}{2} OA \cdot OC \sin COA$$

$$= \frac{1}{2} r_1 r_3 \sin(\theta_3 - \theta_1)$$

Area of $\Delta\, AOB$

$$= \frac{1}{2} OA \cdot OB \sin AOB$$

$$= \frac{1}{2} r_1 r_2 \sin(\theta_1 - \theta_2)$$

So, area of $\Delta\, ABC$

$$= \frac{1}{2} r_2 r_3 \sin(\theta_3 - \theta_2) - \frac{1}{2} r_1 r_3 \sin(\theta_3 - \theta_1) - \frac{1}{2} r_1 r_2 \sin(\theta_1 - \theta_2)$$

$$= \frac{1}{2} r_2 r_3 \sin(\theta_3 - \theta_2) + \frac{1}{2} r_1 r_3 \sin(\theta_1 - \theta_3) + \frac{1}{2} r_1 r_2 \sin(\theta_2 - \theta_1)$$

$$= \frac{1}{2} [r_2 r_3 \sin(\theta_3 - \theta_2) + r_1 r_2 \sin(\theta_2 - \theta_1) + r_1 r_3 \sin(\theta_1 - \theta_3)]$$

ILLUSTRATIVE EXAMPLES

Example 1. Find the polar coordinates of the point whose Cartesian coordinates are $\left(-1, \sqrt{3}\right)$.

Solution : Here $x = -1, y = \sqrt{3}$.

$$\therefore \quad r = \sqrt{x^2 + y^2} = \sqrt{1+3} = 2$$

and $$\tan\theta = \frac{y}{x} = -\sqrt{3} = \tan\frac{2\pi}{3}$$

$$\theta = \frac{2\pi}{3}.$$

Hence, the corresponding polar coordinates are $\left(2, \frac{2\pi}{3}\right)$.

Example 2. Find the cartesian coordinates of the point whose polar coordinates are $(2\sqrt{2}, \pi/4)$.

Solution : Here $r = 2\sqrt{2}$ and $\theta = \pi/4$

$$x = r\cos\theta$$

$$= 2\sqrt{2}\cos\frac{\pi}{4} = 2\sqrt{2}\cdot\frac{1}{\sqrt{2}} = 2$$

$$y = r\sin\theta$$

$$= 2\sqrt{2}\sin\frac{\pi}{4} = 2\sqrt{2}\cdot\frac{1}{\sqrt{2}} = 2$$

Hence, the corresponding cartesian coordinates are (2, 2).

Example 3. Write the polar equation of $y = x\tan\beta$. What does the changed equation represent?

Solution : $y = x\tan\beta$

$$\Rightarrow \quad r\sin\theta = r\cos\theta\tan\beta$$

$$\Rightarrow \quad \tan\theta = \tan\beta$$

$$\Rightarrow \quad \theta = \beta$$

It represents a straight line passing through the pole and making an angle β with the initial line.

Example 4. Find the distance between the points whose polar coordinates are (2, 40°) and (4, 100°).

Solution : We know that the distance between two points (r_1, θ_1) and (r_2, θ_2) is

$$= \sqrt{r_1^2 + r_2^2 - 2r_1r_2 \cos(\theta_2 - \theta_1)}$$

∴ Hence, distance between (2, 40°) and (4, 100°) is,

$$= \sqrt{4 + 16 - 2 \times 2 \times 4 \cos(100° - 40°)}$$

$$= \sqrt{4 + 16 - 16 \cos 60°}$$

$$= \sqrt{4 + 16 - 8} = \sqrt{12} = 2\sqrt{3} \text{ units.}$$

Example 5. Write $\frac{l}{r} = 1 + e \cos \theta$ in cartesian form.

Solution : $$\frac{l}{r} = 1 + e \cos \theta$$

$$\Rightarrow \quad l = r + er \cos \theta$$

$$\Rightarrow \quad l = \sqrt{x^2 + y^2} + ex$$

$$\Rightarrow \quad (l - ex)^2 = x^2 + y^2$$

Example 6. Find the length and direction of the perpendicular drawn from the pole on the line

$$\frac{1}{r} = 8 \cos \theta + 6 \sin \theta.$$

Solution : $$\frac{1}{r} = 8 \cos \theta + 6 \sin \theta$$

$$\Rightarrow \quad \frac{1}{r} = k \cos \alpha \cos \theta + k \sin \alpha \sin \theta$$

where $k \cos \alpha = 8, k \sin \alpha = 6$

Squaring and adding

$$k^2 = 100$$

$$\therefore \quad k = 10$$

Dividing

$$\tan \alpha = \frac{6}{8} = \frac{3}{4}$$

$$\Rightarrow \qquad \frac{1}{r} = k \cos(\theta - \alpha)$$

$$\Rightarrow \qquad \frac{1}{r} = 10 \cos\left(\theta - \tan^{-1}\frac{3}{4}\right)$$

$$\Rightarrow \qquad r \cos\left(\theta - \tan^{-1}\frac{3}{4}\right) = \frac{1}{10}$$

Compare with $r \cos(\theta - \alpha) = p$

We have, $p = \dfrac{1}{10}$ and $\alpha = \tan^{-1}\dfrac{3}{4}$

So, perpendicular from pole $= p = \dfrac{1}{10}$

Inclination of perpendicular from initial line $= \alpha = \tan^{-1}\dfrac{3}{4}$.

Example 7. Find the area of the triangle with vertices

$$(2a, 30°), (4a, 60°)\ (6a, 90°).$$

Solution : Here, $\quad r_1 = 2a, r_2 = 4a, r_3 = 6a$

$\qquad\qquad \theta_1 = 30°, \theta_2 = 60°, \theta_3 = 90°$

$\therefore$ Area of the triangle

$$= \frac{1}{2} r_1 r_2 \sin(\theta_2 - \theta_1) + \frac{1}{2} r_2 r_3 \sin(\theta_3 - \theta_2) + \frac{1}{2} r_3 r_1 \sin(\theta_1 - \theta_3)$$

$$= \frac{1}{2}[(2a)(4a) \sin(60° - 30°) + (4a)(6a) \sin(90° - 60°) + (6a)(2a) \sin(30° - 90°)]$$

$$= \frac{a^2}{2}[8 \sin 30° + 24 \sin 30° - 12 \sin 60°]$$

$$= \frac{a^2}{2}\left[8 \times \frac{1}{2} + 24 \times \frac{1}{2} - 12 \times \frac{\sqrt{3}}{2}\right]$$

$$= \frac{a^2}{2}\left[4 + 12 - 6\sqrt{3}\right] = a^2\left[8 - 3\sqrt{3}\right]$$

Example 8. Prove that the equation

$$m(x^3 - 3xy^2) + y^3 - 3x^2y = 0$$

represents three straight lines equally inclined to one another?

Solution : The given equation is,

$$m\,(x^3 - 3xy^2) + y^3 - 3x^2y = 0.$$

Transforming this equation into polar coordinates, we get

[Put $x = r\cos\theta$, and $y = r\sin\theta$]

$$m\,(r^3\cos^3\theta - 3r\cos\theta \cdot r^2\sin^2\theta) + r^3\sin^3\theta - 3r^2\cos^2\theta \cdot r\sin\theta = 0$$

$$\Rightarrow m\,(\cos^3\theta - 3\cos\theta\sin^2\theta) + \sin^3\theta - 3\cos^2\theta\sin\theta = 0$$

$$\Rightarrow m\,(1 - 3\tan^2\theta) + \tan^3\theta - 3\tan\theta = 0$$

[dividing throughout by $\cos^3\theta$]

$$\Rightarrow \qquad m = \frac{3\tan\theta - \tan^3\theta}{1 - 3\tan^2\theta} = \tan 3\theta$$

If $m = \tan\alpha$, then we have

$$\tan\alpha = \tan 3\theta$$

$$\Rightarrow \quad \tan 3\theta = \tan\alpha$$

$$\Rightarrow \quad 3\theta = n\pi + \alpha$$

$$\Rightarrow \quad \theta = \frac{n\pi}{3} + \frac{\alpha}{3}$$

Put $n = 0, 1, 2$ we find that the given equation represents three straight lines through the origin inclined at angles

$$\frac{\alpha}{3}, 60° + \frac{\alpha}{3}, 120° + \frac{\alpha}{3}$$

to the axis of x.

Hence, the lines are equally inclined to one another.

Example 9. Find the polar coordinates of the foot of the perpendicular from the pole on the line joining the two points (r_1, θ_1) and (r_2, θ_2).

Solution : Let the given straight line be

$$r\cos(\theta - \alpha) = p$$

$$\Rightarrow r\cos\theta\cos\alpha + r\sin\theta\sin\alpha - p = 0$$

It passes through two points (r_1, θ_1) and (r_2, θ_2).

$$\therefore r_1\cos\theta_1\cos\alpha + r_1\sin\theta_1\sin\alpha - p = 0$$

and $$r_2\cos\theta_2\cos\alpha + r_2\sin\theta_2\sin\alpha - p = 0$$

On solving these equations, we get

$$\frac{\cos\alpha}{-r_1\sin\theta_1 + r_2\sin\theta_2} = \frac{\sin\alpha}{-r_2\cos\theta_2 + r_1\cos\theta_1}$$

$$= \frac{p}{r_1 r_2 (\cos\theta_1 \sin\theta_2 - \cos\theta_2 \sin\theta_1)}$$

$$\Rightarrow \quad \frac{\cos\alpha}{r_2 \sin\theta_2 - r_1 \sin\theta_1} = \frac{\sin\alpha}{r_1 \cos\theta_1 - r_2 \cos\theta_2}$$

$$= \frac{p}{r_1 r_2 \sin(\theta_2 - \theta_1)}$$

$$= \frac{\sqrt{\cos^2\alpha + \sin^2\alpha}}{\sqrt{\left[(r_2 \sin\theta_2 - r_1 \sin\theta_1)^2 + (r_1 \cos\theta_1 - r_2 \cos\theta_2)^2\right]}}$$

$$= \frac{1}{\sqrt{r_1^2 + r^2 - 2r_1 r_2 \cos(\theta_2 - \theta_1)}}$$

$$\therefore \quad p = \frac{r_1 r_2 \sin(\theta_2 - \theta_1)}{\sqrt{r_1^2 + r_2^2 - 2r_1 r_2 \cos(\theta_2 - \theta_1)}}$$

and $$\tan\alpha = \frac{r_1 \cos\theta_1 - r_2 \cos\theta_2}{r_2 \sin\theta_2 - r_1 \sin\theta_1}$$

Hence, the coordinates of the foot of the perpendicular are,

$$\left[\frac{r_1 r_2 \sin(\theta_2 - \theta_1)}{\sqrt{r_1^2 + r_2^2 - 2r_1 r_2 \cos(\theta_2 - \theta_1)}}, \tan^{-1}\left(\frac{r_1 \cos\theta_1 - r_2 \cos\theta_2}{r_2 \sin\theta_2 - r_1 \sin\theta_1}\right)\right]$$

Example 10. Find the equation of straight line passing through points whose polar coordinates are (3, 30°) and (4, 90°).

Solution : We know equation of straight line passing through two points whose polar coordinates are (r_1, θ_1) and (r_2, θ_2) is,

$$\frac{\sin(\theta_2 - \theta_1)}{r} = \frac{\sin(\theta - \theta_1)}{r_2} + \frac{\sin(\theta_1 - \theta)}{r_1}$$

So, line through two points whose coordinates are (3, 30°), (4, 90°) is

$$\frac{\sin(90° - 30°)}{r} = \frac{\sin(\theta - 30°)}{4} + \frac{\sin(90° - \theta)}{3}$$

$$\Rightarrow \quad \frac{\sin 60°}{r} = \frac{\sin(\theta - 30°)}{4} + \frac{\cos\theta}{3}$$

$$\Rightarrow \quad \frac{\sqrt{3}}{2r} = \frac{\sin\theta \cos 30° - \cos\theta \sin 30°}{4} + \frac{\cos\theta}{3}$$

$$\Rightarrow \quad \frac{\sqrt{3}}{2r} = \frac{\frac{\sqrt{3}}{2}\sin\theta - \frac{\cos\theta}{2}}{4} + \frac{\cos\theta}{3}$$

$$\Rightarrow \quad \frac{\sqrt{3}}{2r} = \frac{\sqrt{3}\sin\theta - \cos\theta}{8} + \frac{\cos\theta}{3}$$

$$\Rightarrow \quad \frac{\sqrt{3}}{r} = \frac{3\sqrt{3}\sin\theta - 3\cos\theta + 8\cos\theta}{12}$$

$$\Rightarrow \quad 12\sqrt{3} = r(3\sqrt{3}\sin\theta + 5\cos\theta)$$

$$\Rightarrow \quad \frac{12\sqrt{3}}{r} = 5\cos\theta + 3\sqrt{3}\sin\theta$$

which is required equation of straight line through given points.

Example 11. A circle passes through the point (r_1, θ_1) and touches the initial line at a distance c from the pole. Show that its polar equation is

$$\frac{r^2 - 2cr\cos\theta + c^2}{r\sin\theta} = \frac{r_1^2 - 2cr_1\cos\theta_1 + c^2}{r_1\sin\theta_1}.$$

Solution : The equation of the circle whose centre is (R, α) and radius is a is given by

$$r^2 - 2Rr\cos(\theta - \alpha) + R^2 - a^2 = 0 \qquad \ldots\text{(i)}$$

Let it touch the initial line at N, such that $ON = c$. Then,

and $$\left.\begin{aligned} a &= c\tan\alpha \\ R &= c\sec\alpha \end{aligned}\right\}$$

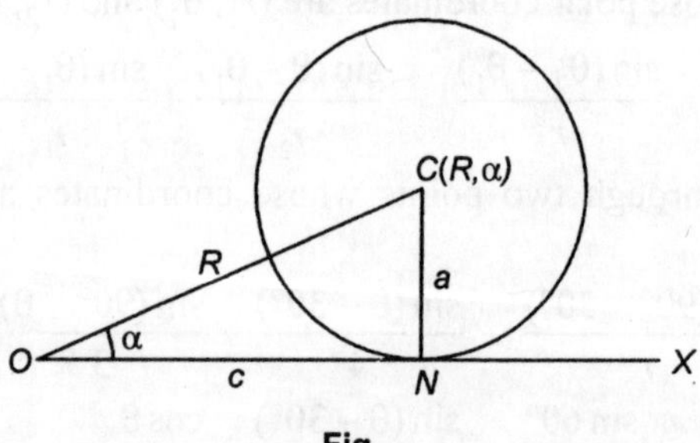

Fig.

Hence, the equation of the circle becomes,

$$r^2 - 2c\sec\alpha\cdot r\cdot\cos(\theta - \alpha) + c^2\sec^2\alpha - c^2\tan^2\alpha = 0$$

$\Rightarrow \quad r^2 - 2cr \sec\alpha\,(\cos\theta\cos\alpha + \sin\theta\sin\alpha) + c^2 = 0$

$\Rightarrow \quad r^2 - 2rc\,(\cos\theta + \sin\theta\tan\alpha) + c^2 = 0$

$$\Rightarrow \quad \frac{r^2 - 2cr\cos\theta + c^2}{r\sin\theta} = 2c\tan\alpha \qquad \text{...(ii)}$$

If it passes through (r_1, θ_1), then we have

$$\frac{r_1^2 - 2cr_1\cos\theta_1 + c^2}{r_1\sin\theta_1} = 2c\tan\alpha \qquad \text{...(iii)}$$

From (ii) and (iii), equating the two values of $2c\tan\alpha$, we have

$$\frac{r^2 - 2cr\cos\theta + c^2}{r\sin\theta} = \frac{r_1^2 - 2cr_1\cos\theta_1 + c^2}{r_1\sin\theta_1}$$

which is the required equation of the circle.

Example 12. Find the condition that the straight line

$$\frac{1}{r} = a\cos\theta + b\sin\theta$$

may touch the circle $r = 2c\cos\theta$.

Solution : Eliminating θ between the two equations, the radii vectors of the points of intersection are given by,

$$\frac{1}{r} = a\cdot\frac{r}{2c} + b\cdot\sqrt{1 - \left(\frac{r}{2c}\right)^2}$$

$$\Rightarrow \quad \left(\frac{2c}{r} - ar\right)^2 = b^2\,(4c^2 - r^2)$$

$\Rightarrow \quad (2c - ar^2)^2 = b^2r^2\,(4c^2 - r^2)$

$\Rightarrow \quad r^4\,(a^2 + b^2) - 4r^2c\,(a + b^2c) + 4c^2 = 0$

This is a quadratic equation in r^2. Hence, it will give two values of r^2. The line will touch the circle, if these two values of r^2 are equal

i.e., if $\quad 16c^2\,(a + b^2c)^2 - 16\,(a^2 + b^2)\,c^2 = 0$

i.e., if $\quad a^2 + b^4c^2 + 2ab^2c - a^2 - b^2 = 0$

i.e., if $\quad b^2c^2 + 2ac - 1 = 0$

i.e., if $\quad b^2c^2 + 2ac = 1$

which is the required condition.

Example 13. Find the equation of circle whose centre is (R, α), radius is a and initial line is one of the tangents of circle.

Solution : Let O be the pole, OX be the initial line, $C\ (R, \alpha)$ centre of circle whose radius is a. Initial line is tangent of circle. CN is perpendicular from C on tangent. Let $P\ (r, \theta)$ is any point on circle.

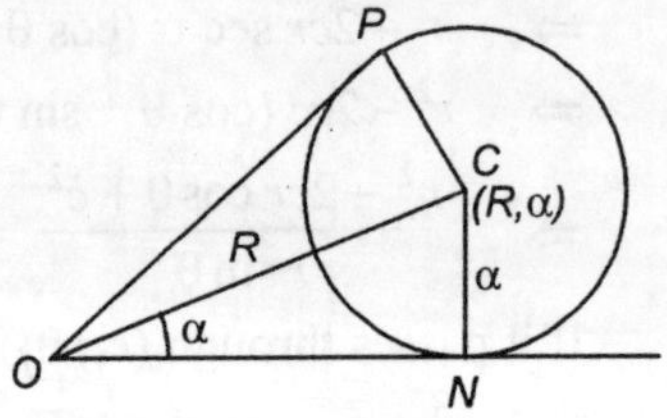

Fig.

$OP = r,\ OC = R,\ \angle PON = \theta,$
$\angle CON = \alpha,\ \angle POC = \theta - \alpha.$

In $\Delta\ POC$

$$(PC)^2 = (OP)^2 + (OC)^2 - 2OP \cdot OC \cos POC$$

$$a^2 = r^2 + R^2 - 2rR\cos(\theta - \alpha) \qquad \ldots(i)$$

In $\qquad \Delta OCN,\ \sin\alpha = \dfrac{a}{R}$

$\therefore \qquad R = a\operatorname{cosec}\alpha \qquad \ldots(ii)$

From (i) and (ii)

$$a^2 = r^2 + a^2\operatorname{cosec}^2\alpha - 2ra\operatorname{cosec}\alpha\cos(\theta - \alpha)$$

$$\Rightarrow \qquad r^2 + a^2\cot^2\alpha - 2ar\operatorname{cosec}\alpha\cos(\theta - \alpha) = 0$$

which is required equation of circle.

Example 14. Find the centre of circle

$$r = A\cos\theta + B\sin\theta.$$

Solution : Equation of circle is,

$$r = A\cos\theta + B\sin\theta \qquad \ldots(i)$$

Let $\qquad A = r_1\cos\beta,\ B = r_1\sin\beta$

then $\qquad r_1 = \sqrt{A^2 + B^2}$

$$\tan\beta = \frac{B}{A}$$

$$\Rightarrow \qquad \beta = \tan^{-1}\frac{B}{A}$$

Equation of circle (i) is,

$$r = r_1\cos\theta\cos\beta + r_1\sin\theta\sin\beta$$

$$= r_1\cos(\theta - \beta) \qquad \ldots(ii)$$

We know that the equation of circle whose centre is (a, α) is,

$$r = 2a\cos(\theta - \alpha) \qquad \ldots(iii)$$

Comparing (ii) and (iii),

$$r_1 = 2a, \beta = \alpha$$

So, polar coordinates of centre of circle (a, α) are

$$\left\{\frac{1}{2}\sqrt{A^2 + B^2}, \tan^{-1}\frac{B}{A}\right\}.$$

Example 15. Find the polar equation of circle whose diameter is a line joining points whose polar coordinates are (a, α), (a, β).

Solution : Let AB be the diameter of a circle such that polar coordinates of A and B are (a, α) and (b, β).

Let $P(r, \theta)$ be any point on circle.

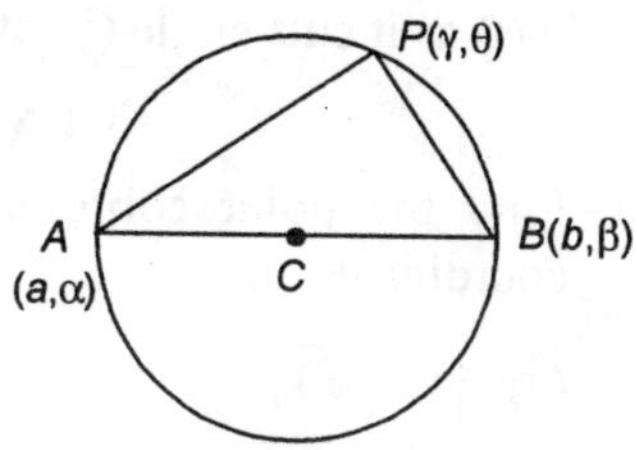

Fig.

Clearly, $\angle APB = \frac{\pi}{2}$

So, $(AB)^2 = (AP)^2 + (BP)^2$ ...(i)

We know distance between two points (r_1, θ_1) and (r_2, θ_2)

$$= \sqrt{r_1^2 + r_2^2 - 2r_1r_2\cos(\theta_2 - \theta_1)}$$

Hence,
$$(AB)^2 = a^2 + b^2 - 2ab\cos(\beta - \alpha) \quad \text{...(ii)}$$
$$(AP)^2 = r^2 + a^2 - 2ar\cos(\theta - \alpha) \quad \text{...(iii)}$$
$$(BP)^2 = r^2 + b^2 - 2br\cos(\theta - \beta) \quad \text{...(iv)}$$

Putting in (i),

$$a^2 + b^2 - 2ab\cos(\beta - \alpha) = a^2 + r^2 - 2ar\cos(\theta - \alpha) + b^2 + r^2 - 2br\cos(\theta - \beta)$$

$$\Rightarrow 2r^2 - 2ar\cos(\theta - \alpha) - 2br\cos(\theta - \beta) + 2ab\cos(\beta - \alpha) = 0$$

Hence, required circle is,

$$r^2 - ar\cos(\theta - \alpha) - br\cos(\theta - \beta) + ab\cos(\beta - \alpha) = 0$$

Example 16. Show that polar equations $r = 8\cos\theta$ and $r\sin\theta = 4$ represent separately circle and its tangent.

Solution:
$$r = 8\cos\theta \quad \text{...(i)}$$

is of form $r = 2a\cos\theta$ so it is clearly the equation of circle

$$r\sin\theta = 4 \quad \text{...(ii)}$$

$\Rightarrow y = 4$ which is clearly a equation of a straight line.

Eliminating θ between (i) and (ii)

$$r^2 = 64\cos^2 \theta$$
$$= 64\{1 - \sin^2\theta\}$$

$$= 64\left\{1 - \frac{16}{r^2}\right\} \qquad \text{[from eqn. (ii)]}$$

$$= \frac{64\{r^2 - 16\}}{r^2}$$

$$\Rightarrow \quad r^4 - 64r^2 + 64 \times 16 = 0$$

$$\Rightarrow \quad (r^2 - 32)^2 = 0 \qquad \text{...(iii)}$$

Clearly two roots of (iii) are equal, so line (ii) is a tangent line of circle (i) as it cuts circle (i) at two coincident points.

EXERCISE 1

1. Find the polar coordinates of the points whose cartesian coordinates are :

(a) $\left(1, -\sqrt{3}\right)$ (b) $\left(-\sqrt{3}, 1\right)$

2. Find the cartesian coordinates of the points whose polar coordinates are :

(a) $(3, \pi/3)$ (b) $(-5, \pi/4)$

3. Transform the following equations to the corresponding polar coordinate equations :

(a) $y^2 + 4ax = 4a^2$ (b) $x^2 + y^2 = 16$

(c) $y^2 = 4ax$ (d) $3x + y = 0$

(e) $x^2 + y^2 - 2x + 2y = 0$ (f) $(x^2 + y^2)^2 = a^2 (x^2 - y^2)$

4. Transform the following equations to the corresponding cartesian coordinate equations :

(a) $r = 4a \cos \theta$ (b) $r^2 = a^2 \cos 2\theta$

(c) $r = \delta \cos \theta$ (d) $r = \dfrac{4}{1 + \cos\theta}$

(e) $\sqrt{r} \cos\dfrac{\theta}{2} = \sqrt{a}$

5. Find the lengths of the straight lines joining the pair of points whose polar coordinates are :

(a) (2, 10°) and (2, 40°) (b) (3, 20°) and (4, 50°)

(c) (2, 40°) and (4, 100°) (d) (3, 60°) and (5, 150°)

(e) (– 6, 30°) and (4, 90°)

6. Prove that the points (0, 0°), (3, 90°) and (3, 30°) form an equilateral triangle.

7. Find the areas of the triangle whose vertices are :

(a) (2, 60°), (3, 90°), (4, 120°)

(b) (– 6, 30°), (4, 90°), (5, 150°)

(c) (9, 0°), $\left(9, \frac{\pi}{2}\right), \left(9, \frac{3\pi}{4}\right)$

(d) $\left(-a, \frac{\pi}{6}\right), \left(a, \frac{\pi}{2}\right), \left(-2a, -\frac{2\pi}{3}\right)$

8. Find the perpendicular distance from pole to the line

$$\frac{6}{r} = \cos\theta + \sqrt{3}\sin\theta.$$

9. Show that the equation $l/r = f(\theta)$

where $\qquad f(\theta) = a\cos(\theta + \alpha) + b\cos(\theta + \beta),$

represents a straight line, and that any straight line perpendicular to it is $\lambda/r = f(\theta + \pi/2)$ where λ is a parameter.

10. Prove that the straight line passing through the point of intersection of the lines

$$\frac{1}{r} = a\cos\theta + b\sin\theta$$

and
$$\frac{1}{r} = a'\cos\theta + b'\sin\theta$$

is
$$\frac{1+\lambda}{r} = (a + \lambda a')\cos\theta + (b + \lambda b')\sin\theta$$

Hence, obtain the equation of the line passing through the pole and the point of intersection of the two given lines.

[***Hint.*** A line passing through the pole does not contain r.]

11. Show that the condition that the chord cut off by the curve $l/r = 1 + e\cos\theta$ from the line,

$$\frac{l}{r} = a\cos\theta + b\sin\theta$$

may subtend a right angle at the pole is,

$$(la - e)^2 + l^2b^2 = 2$$

12. Show that the circles $r = 2a\cos(\theta - \alpha)$ and $r = 2a\cos(\theta - \beta)$ intersect at an angle $\alpha - \beta$.

13. Prove that the equations

$$r = a\cos(\theta - \alpha)$$

and
$$r = b\sin(\theta - \alpha)$$

represent two circles which intersect orthogonally.

14. Show that for all values of a, the straight line

$$r\cos(\theta-\alpha)=a+r_1\cos\alpha,$$

is a tangent to the circle

$$r^2-2rr_1\cos\theta+r_1^2-a^2=0.$$

15. Prove that the equation

$$r^2\cos\theta-ar\cos 2\theta-2a^2\cos\theta=0$$

represents a circle and a straight line.

16. Find the perpendicular from pole to the line $\frac{l}{r}=4\cos\theta-3\sin\theta$. Also, find the angle which perpendicular makes with initial line.

ANSWERS
Exercise 1

1. (a) $\left(2,\frac{2\pi}{3}\right)$ (b) $\left(2,\frac{5\pi}{6}\right)$

2. (a) $\left(\frac{3}{2},\frac{3\sqrt{3}}{2}\right)$ (b) $\left(-\frac{5}{\sqrt{2}},-\frac{5}{\sqrt{2}}\right)$

3. (a) $r^2\sin^2\theta+4ar\cos\theta=4a^2$ (b) $r=4$
(c) $r^2\sin^2\theta=4ar\cos\theta$ (d) $r(3\cos\theta+\sin\theta)=0$
(e) $r-2\cos\theta+2\sin\theta=0$ (f) $r^2=a^2\cos 2\theta$

4. (a) $x^2+y^2=4ax$ (b) $(x^2+y^2)^2=a^2(x^2-y^2)$
(c) $x^2+y^2=\delta x$ (d) $x^2+y^2=(4-x)^2$
(e) $y^2=4a^2-4ax$

5. (a) $2\sqrt{2-\sqrt{3}}$ (b) $\sqrt{25-12\sqrt{3}}$ (c) $2\sqrt{3}$
(d) $\sqrt{34}$ (e) $\sqrt{76}$

7. (a) $\frac{9}{2}-2\sqrt{3}$ (b) $\frac{13\sqrt{3}}{2}$ (c) $\frac{a^2}{2}$ (d) $\frac{\sqrt{3}a^2}{4}$

8. $\frac{3\sqrt{7}}{7}$ **10.** $\tan\theta=\frac{a'-a}{b-b'}$ **16.** $\frac{1}{5}$, $\tan^{-1}\left(-\frac{3}{4}\right)$

OBJECTIVE TYPE EXAMPLES

Example 1. Area of triangle whose vertices are $(a, 0)$, $\left(a, \frac{\pi}{2}\right)$ and $\left(a, \frac{3\pi}{4}\right)$ is

(i) a^2 (ii) $\frac{a^2}{2}$ (iii) $\frac{3}{2}a^2$ (iv) $2a^2$

Solution : From 1.11 we know area of triangle whose vertices are (r_1, θ_1), (r_2, θ_2) and (r_3, θ_3) is,

$$= \frac{1}{2}[r_1 r_2 \sin(\theta_2 - \theta_1) + r_2 r_3 \sin(\theta_3 - \theta_2) + r_1 r_3 \sin(\theta_1 - \theta_3)]$$

Here $r_1 = a, r_2 = a, r_3 = a, \theta_1 = 0, \theta_2 = \frac{\pi}{2}, \theta_3 = \frac{3\pi}{4}$

So, area of Δ is,

$$= \frac{1}{2}\left[a^2 \sin\left(\frac{\pi}{2} - 0\right) + a^2 \sin\left(\frac{3\pi}{4} - \frac{\pi}{2}\right) + a^2 \sin\left(0 - \frac{3\pi}{4}\right)\right]$$

$$= \frac{1}{2}\left[a^2 \sin\frac{\pi}{2} + a^2 \sin\frac{\pi}{4} - a^2 \sin\frac{3\pi}{4}\right]$$

$$= \frac{a^2}{2}\left[1 + \frac{1}{\sqrt{2}} - \frac{1}{\sqrt{2}}\right] = \frac{a^2}{2}$$

So, choice (ii) is correct.

OBJECTIVE EXERCISE

1. Polar equation of line joining (a, α) and (b, β) is,

(i) $\frac{\sin(\beta - \alpha)}{r} = \frac{\sin(\theta - \alpha)}{b} + \frac{\sin(\theta - \beta)}{a}$

(ii) $\frac{\sin(\theta - \beta)}{r} = \frac{\sin(\theta - \alpha)}{a} + \frac{\sin(\beta - \theta)}{b}$

(iii) $\frac{\sin(\beta - \alpha)}{r} = \frac{\sin(\alpha - \theta)}{b} + \frac{\sin(\theta - \beta)}{a}$

(iv) None of these.

2. Distance between points (a, α), $(a, \pi + \alpha)$ is,

(i) a (ii) $2a$

(iii) 0 (iv) $\sqrt{2}a$.

3. Equation $r = a$ represents one :

(i) straight line (ii) circle
(iii) cone (iv) cylinder.

4. A straight line passes through pole and makes an angle β with initial line. Its equation is,

(i) $r \cos(\theta - \beta) = p$ (ii) $r \sin(\theta - \beta) = p$
(iii) $\theta = \beta$ (iv) None of these.

5. Equation $r \cos(\theta - \alpha) = p$ represents :

(i) circle (ii) ellipse
(iii) parabola (iv) straight line.

6. General equation of straight line in polar system is,

(i) $r \cos(\theta - \alpha) = p$ (ii) $r \sin(\theta - \alpha) = p$
(iii) $r \sin\theta = p$ (iv) $r \cos\theta = p$.

7. Polar equation of a circle passing through pole is,

(i) $r = a$ (ii) $r = 2a \cos\theta$
(iii) $r = 2a \cos(\theta - \alpha) + c$ (iv) $r = 2a \cos(\theta - \alpha)$.

8. Polar equation of $y = x \tan\beta$ is,

(i) $\theta = \beta$ (ii) $r \sin\theta = \tan\beta$
(iii) $r \sin\theta = \sin\beta$ (iv) None of these.

9. Equation of circle whose centre is pole and radius is a,

(i) $r = a$ (ii) $r = 2a \cos\theta$
(iii) $r = 2a \sin\theta$ (iv) None of above.

10. $\theta = \beta$ represents,

(i) a constant angle (ii) circle
(iii) conic (iv) straight line.

11. Distance between two points (r_1, α) and (r_2, β) is,

(i) $\sqrt{r_1^2 + r_2^2 + 2r_1r_2 \cos(\beta - \alpha)}$

(ii) $\sqrt{r_1^2 + r_2^2 - 2r_1r_2 \cos(\alpha - \beta)}$

(iii) $\sqrt{r_1^2 + r_2^2 + 2r_1r_2 \cos(\alpha + \beta)}$

(iv) $\sqrt{r_1^2 + r_2^2 - 2r_1r_2 \cos(\alpha + \beta)}$

12. Cartesian form of polar equation $r^{1/2} = a^{1/2} \cos\frac{\theta}{2}$ is,

(i) $2x^2 + 2y^2 - ax^2 = a^2 (x^2 + y^2)$

(ii) $2x^2 + 2y^2 + ax^2 = a^2 (x^2 + y^2)$

(iii) $2x^2 + 2y^2 - ax^2 = a^2 (x^2 - y^2)$

(iv) None of above.

ANSWERS

Objective Exercise

1. (iv)	**2.** (ii)	**3.** (ii)	**4.** (i)	**5.** (iv)
6. (i)	**7.** (ii)	**8.** (i)	**9.** (i)	**10.** (iv)
11. (ii)	**12.** (iv)			

2

Polar Equations of Conics

2.1 Polar Equation of a Conic

To find the polar equation of a conic, the focus being the pole.

Let S be the focus, A the vertex, and ZM the directrix of the conic. Let e, be the eccentricity of the conic.

Draw SZ perpendicular to the directrix ZM.

Let S be the pole and SZ be initial line.

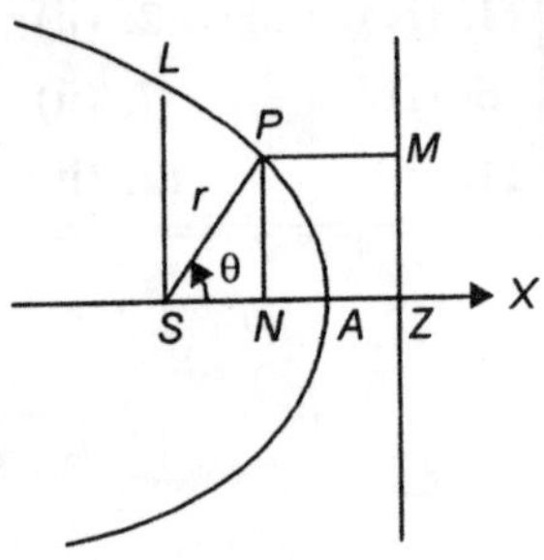

Fig.

Take any point P on the conic. Let its polar coordinates be (r, θ).

Then, $\qquad SP = r$

or $\qquad \angle ZSP = \theta$

Draw PM, PN perpendiculars respectively to the directrix ZM and the initial line SZX.

Let SL be the semi-latus rectum, and let $SL = l$.

Then by definition of a conic, we have

$$SP = e \cdot PM$$

$$\Rightarrow \quad SP = e\,(NZ)$$

$$\Rightarrow \quad SP = e\,(SZ - SN) \qquad \ldots(1)$$

Also, $$\frac{SL}{SZ} = e$$

$$\Rightarrow \quad SZ = \frac{SL}{e} = \frac{l}{e}$$

$\therefore$ From (1),

$$SP = e\left(\frac{l}{e} - SN\right)$$

$$\Rightarrow \quad r = e\left(\frac{l}{e} - r\cos\theta\right)$$

$$\Rightarrow \quad r = l - er\cos\theta$$

$$\Rightarrow \quad l = r(1 + e\cos\theta)$$

$$\Rightarrow \quad \boxed{\frac{l}{r} = 1 + e\cos\theta} \quad \ldots(2)$$

Since, (2) is the relation between the polar coordinates of *any* point on the conic, hence it is the required equation of the conic.

Corollary 1. If the line *ZS* (produced) is taken as the positive direction of the initial line, we have

$$\frac{SL}{SZ} = e$$

$$\Rightarrow \quad \frac{l}{SZ} = e$$

$$\Rightarrow \quad SZ = \frac{l}{e}$$

Also, $SP = e \cdot PM$

$$= e \cdot NZ$$

$$= e(ZS + SN)$$

$$\Rightarrow \quad r = e\left(\frac{l}{e} + r\cos\theta\right)$$

$$\Rightarrow \quad r = l + er\cos\theta$$

$$\Rightarrow \quad l = r(1 - e\cos\theta)$$

$$\Rightarrow \quad \boxed{\frac{l}{r} = 1 - e\cos\theta} \quad \ldots(3)$$

Corollary 2. If the given conic is a parabola, then $e = 1$.

$\therefore$ Equation (2) becomes

$$\frac{l}{r} = 1 + \cos\theta$$

$$\Rightarrow \quad \boxed{\frac{l}{r} = 2\cos^2\frac{\theta}{2}} \quad \ldots(4)$$

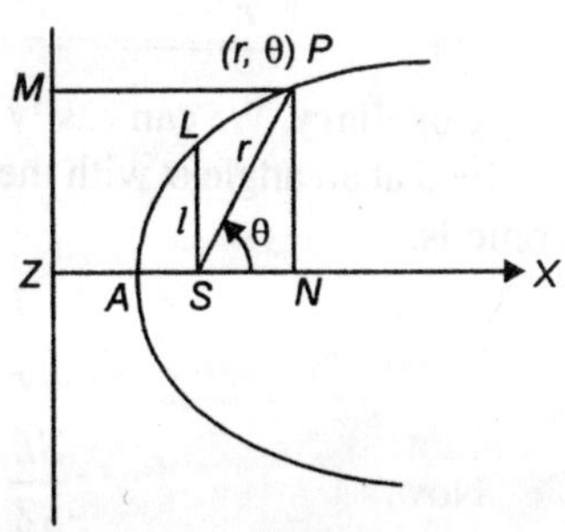

Fig.

and equation (3) becomes

$$\frac{l}{r} = 1 - \cos\theta$$

$$\Rightarrow \quad \boxed{\frac{l}{r} = 2\sin^2\frac{\theta}{2}} \qquad \ldots(5)$$

2.2 Axis Inclined at an Angle α

To find the polar equation of a conic with its focus as pole and its axis inclined at angle α to initial line.

Let the axis SZ of the conic makes an angle α with the initial line SX. Then,

$$\angle PSZ = \theta - \alpha \text{ where } P \to (r, \theta)$$

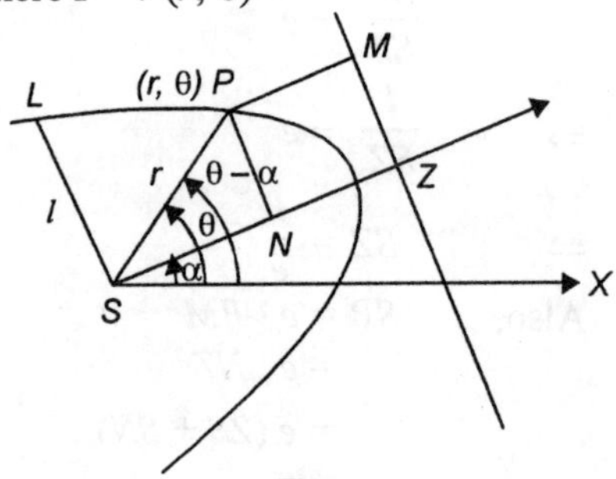

Fig.

Now $\dfrac{SL}{SZ} = e$

$\Rightarrow \quad \dfrac{l}{SZ} = e$

$\Rightarrow \quad SZ = \dfrac{l}{e}$

Also, $SP = e \cdot PM$

$\Rightarrow \quad SP = e \cdot NZ$

$\Rightarrow \quad SP = e\,(SZ - SN)$

$\Rightarrow \quad r = e\,[l/e - r\cos(\theta - \alpha)]$

$\Rightarrow \quad r = l - er\cos(\theta - \alpha)$

$\Rightarrow \quad l = r\,\{1 + e\cos(\theta - \alpha)\}$

$$\Rightarrow \quad \boxed{\frac{l}{r} = l + e\cos(\theta - \alpha)} \qquad \ldots(1)$$

Corollary. We can easily show that if the axis ZS of the conic be inclined at an angle α with the initial line SX, then the equation of the conic is,

$$\frac{l}{r} = 1 - e\cos(\theta - \alpha)$$

Now, $\dfrac{SL}{SZ} = e \Rightarrow \dfrac{l}{SZ} = e \Rightarrow SZ = \dfrac{l}{e}$

Also, $SP = e \cdot PM = e \cdot NZ = e\,(SZ + SN)$

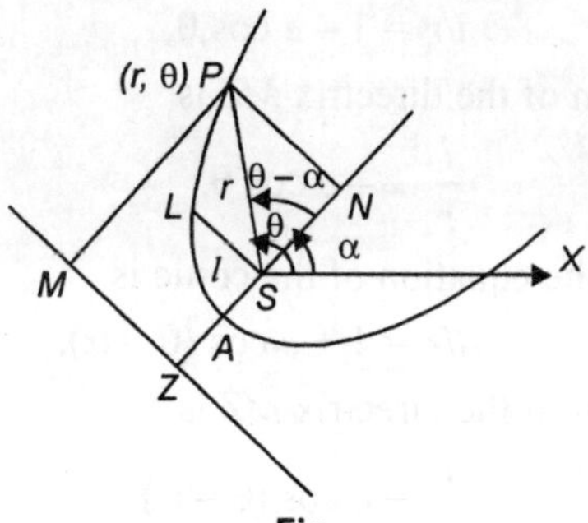

Fig.

$$\Rightarrow \quad r = e\,[l/e + r\cos(\theta - \alpha)]$$

$$\Rightarrow \quad r = l + er\cos(\theta - \alpha)$$

$$\Rightarrow \quad l = r\,\{1 - e\cos(\theta - \alpha)\}$$

$$\Rightarrow \quad \boxed{\frac{l}{r} = 1 - e\cos(\theta - \alpha)} \qquad \ldots(2)$$

2.3 Equation to the Directrices

(A) *To find the polar equation of the directrix of the conic $l/r = 1 + e\cos\theta$ corresponding to the focus which is the pole.*

We are to find out the equation of the directrix *MZ*.

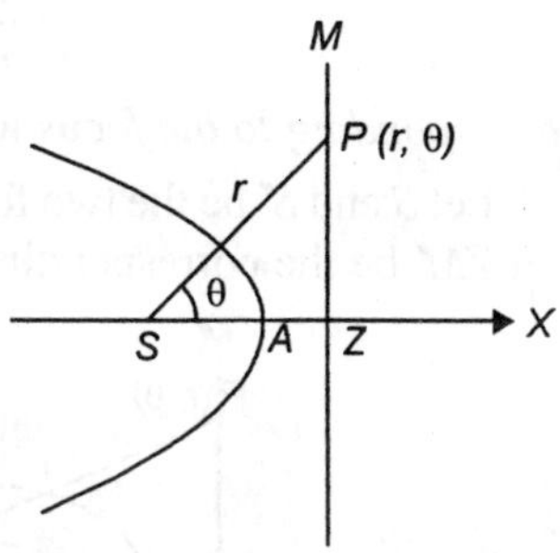

Fig.

Let (r, θ) be the polar coordinates of any point *P* on the directrix *MZ*. Then $SP = r$ and $\angle PSZ = \theta$.

$$\therefore \quad SZ = SP\cos\theta$$

$$\Rightarrow \quad \frac{l}{e} = r\cos\theta$$

$$\Rightarrow \quad \boxed{\frac{l}{r} = e\cos\theta} \qquad \ldots(1)$$

This is the relation between the polar coordinates of *any* point on the directrix *MZ*, hence is the equation of the directrix *MZ*.

Corollary 1. If the equation of the conic is

$$l/r = 1 - e\cos\theta,$$

then the equation of the directrix MZ is

$$\frac{l}{r} = -e\cos\theta.$$

Corollary 2. If the equation of the conic is

$$l/r = 1 + e\cos(\theta - \alpha),$$

then the equation of the directrix MZ is

$$\frac{l}{r} = e\cos(\theta - \alpha).$$

Corollary 3. If the equation of the conic is

$$l/r = 1 - e\cos(\theta - \alpha),$$

then the equation of the directrix MZ is

$$\frac{l}{r} = -e\cos(\theta - \alpha)$$

(B) *To find the polar equation of the directrix of the conic*

$$\frac{l}{r} = 1 + e\cos\theta$$

corresponding to the focus which is not the pole.

Let S and S' be the two foci of the conic $l/r = 1 + e\cos\theta$. Let ZM and ZM' be the corresponding directrices respectively.

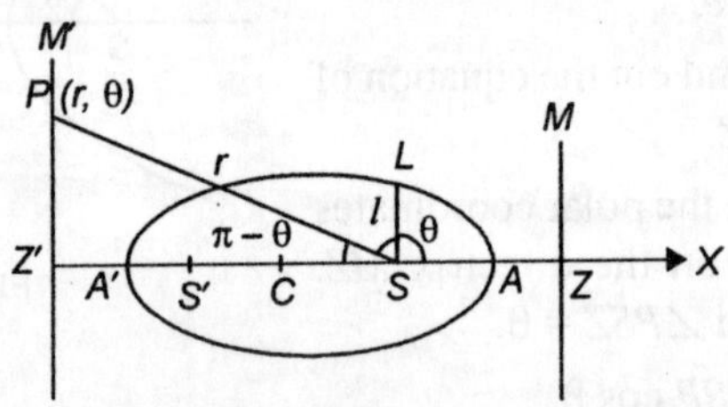

Fig.

Here one of the foci S is the pole and SZ is the initial line.

We are to find out the equation of the directrix $M'Z'$.

Let $P(r, \theta)$ be any point on $M'Z'$. Then,

$$SZ' = SP\cos(\pi - \theta) = r\cos(\pi - \theta)$$
$$= -r\cos\theta$$

$$\Rightarrow \quad ZZ' - SZ = -r\cos\theta$$

$$\Rightarrow \quad \frac{2a}{e} - \frac{l}{e} = -r\cos\theta$$

$$\left[\because ZZ' = 2CZ' = 2\cdot\frac{a}{e} \text{ and } SZ = \frac{l}{e}\right]$$

$$\Rightarrow \quad \frac{2a}{e} - \frac{b^2}{ae} = -r\cos\theta \qquad \left[\because l = \frac{b^2}{a}\right]$$

$$\Rightarrow \quad \frac{2a}{e} - \frac{a^2(1-e^2)}{ae} = -r\cos\theta \qquad [\because b^2 = a^2(1-e^2)]$$

$$\Rightarrow \quad \frac{2a}{e} - \frac{a^2(1-e^2)}{e} = -r\cos\theta$$

$$\Rightarrow \quad a[2 - 1 + e^2] = -er\cos\theta$$

$$\Rightarrow \quad a(1 + e^2) = -er\cos\theta$$

Now,
$$l = \frac{b^2}{a^2} = \frac{a^2(1-e^2)}{a}$$
$$= a(1-e^2)$$

$$\therefore \quad a = \frac{l}{1-e^2}$$

$$\therefore \quad \frac{l(1+e^2)}{1-e^2} = -er\cos\theta$$

$$\Rightarrow \quad \boxed{\frac{l}{r} = -\left(\frac{1-e^2}{1+e^2}\right)e\cos\theta}$$

which is the required equation.

Note 1. If the equation of the conic be $l/r = 1 - e\cos\theta$, then the corresponding equation of the other directrix will be

$$\boxed{\frac{l}{r} = \left(\frac{1-e^2}{1+e^2}\right)\cdot e\cos\theta}$$

Note 2. We have derived above the equations of the other directrix in case of ellipse ($e < 1$). We can easily show that the same equations hold good in case of hyperbola also ($e > 1$).

2.4 Tracing of the Conic

$$l/r = 1 + e \cos \theta.$$

Case I. *When the conic is a parabola. In this case, e = 1. Therefore, the equation of the conic becomes l/r = 1 + cos θ.*

Since, cos (– θ) = cos θ, the equation of the conic remains unaltered when θ is replaced by – θ. Therefore, the conic is symmetrical about the initial line.

Consequently, it is sufficient to consider the variation of θ in first and second quadrants only, i.e., from $0 \le \theta \le \pi$.

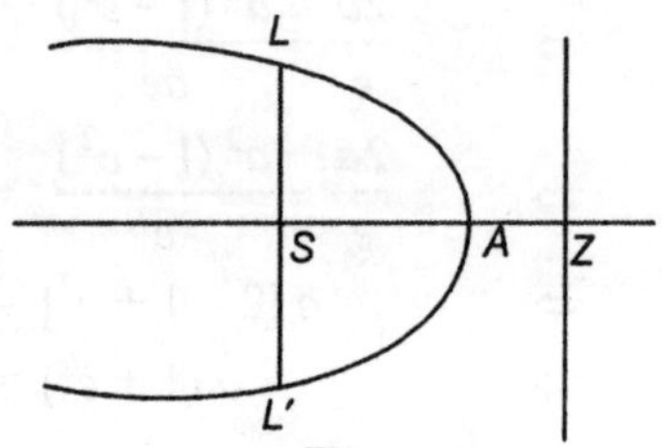

Fig.

When $\theta = 0$, $\cos \theta = 1$, $\dfrac{l}{r} = 2$

$$\Rightarrow r = \frac{l}{2}$$

This corresponds to the point A, the vertex where $r = SA = l/2$.

When θ increases from 0 to π/2, cos θ decreases from 1 to 0 so that l/r decreases from 2 to 1. This corresponds to the part of the curve between A and L.

When $\qquad \theta = \dfrac{\pi}{2}, \cos \theta = 0$

so that $\qquad \dfrac{l}{r} = 1 \Rightarrow r = l$

This corresponds to the point L, the end point of the latus rectum. When θ increases from π/2 to π, cos θ decreases from 0 to –1, so that r increases from l to infinity. This corresponds to the part of the curve ahead to L.

When $\theta = \pi$, $\cos \theta = -1$, $l/r = 0 \Rightarrow r = \infty$

The curve from θ = π to θ = 2π, i.e., the portion of the parabola below the initial line can be traced by symmetry.

Case II. *When the conic is an ellipse (e < 1).*

$$\frac{l}{r} = 1 + e \cos \theta$$

$$\Rightarrow \qquad r = \frac{l}{1 + e \cos \theta}$$

When $\theta = 0$, $r = \dfrac{l}{1+e}$, i.e., the curve meets the axis at the point A such that $r = SA = \dfrac{l}{1+e}$. It is to be noted here that since the maximum

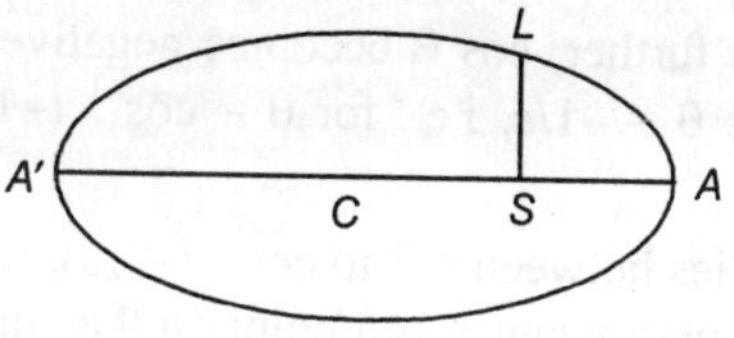

Fig.

value of $\cos\theta$ is 1, hence at A the value of r is minimum. When θ increases from 0 to $\pi/2$, $\cos\theta$ decreases from 1 to 0, so that $1 + e\cos\theta$ decreases from $1 + e$ to 1, i.e., r increases from $\dfrac{l}{1+e}$ to l. This corresponds to the portion AL of the curve. At L, $\theta = \pi/2$ $\therefore r = l$. This corresponds to the end-point of the latus-rectum.

When θ increases from $\pi/2$ to π, $\cos\theta$ decreases from 0 to -1, so that $1 + e\cos\theta$ decreases from 1 to $(1 - e)$, i.e., r increases from l to $\dfrac{l}{1-e}$. This corresponds to the portion of the curve ahead to L.

When $\theta = \pi$, $r = \dfrac{l}{1-e}$. This corresponds to the point A'. Since, for $\theta = \pi$, $\cos\theta = -1$, which is the minimum value of $\cos\theta$, hence, r is greatest at A'.

The curve from $\theta = \pi$ to $\theta = 2\pi$, i.e., the portion of the ellipse below the initial line can be traced by symmetry.

Case III. *When the conic is a hyperbola ($e > 1$) :*

$$\frac{l}{r} = 1 + e\cos\theta$$

$$\Rightarrow \qquad r = \frac{l}{1 + e\cos\theta}$$

When $\theta = 0$, $r = \dfrac{l}{1+e}$, i.e., the curve meets the axis at the point A such that $SA = \dfrac{l}{1+e}$.

It is to be noted here that since the maximum value of $\cos\theta$ is 1, hence at A, the value of r is minimum.

As θ increases, $\cos\theta$ decreases and therefore, r increases. When $\theta = \pi/2$, $r = l$. This corresponds to the point L. Thus, as θ increases from 0 to $\pi/2$, r increases from $\dfrac{l}{1+e}$ to l. This corresponds to the part of the curve between A and L.

As θ increases further, $\cos\theta$ becomes negative and r goes on increasing. For $\cos\theta = -1/e$, i.e., for $\theta = \cos^{-1}(-1/e)$, r becomes infinite.

Thus, when θ lies between $\pi/2$ to $\cos^{-1}(-1/e)$, it corresponds to the part LK of the curve when K is a point on the curve at infinity.

If θ further increases beyond $\cos^{-1}(-1/e)$, r becomes negative and this negative value will continue till $\theta = \pi$.

When $\theta = \pi$, $r = \dfrac{l}{1+e}$ i.e., $\dfrac{l}{1+e}$ as $e > 1$. This corresponds to the point A' on the further branch. Hence, the value of θ from $\cos^{-1}(-1/e)$ to π corresponds to the part $A'M$, where M is on the curve at infinity. For values of θ between π and 2π, the curve can be traced by symmetry.

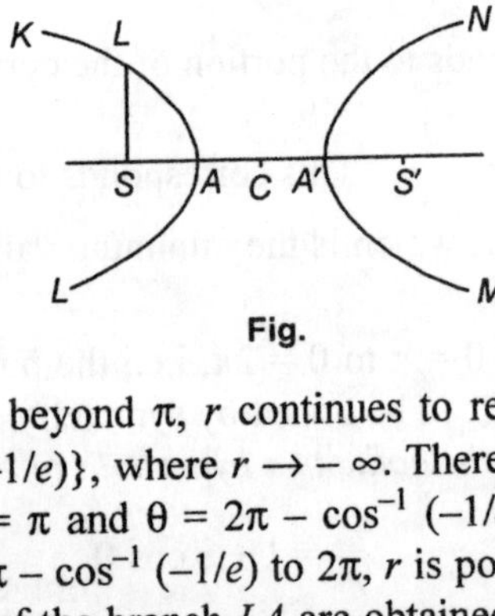

Fig.

If θ increases beyond π, r continues to remain negative until $\theta = \{2\pi - \cos^{-1}(-1/e)\}$, where $r \to -\infty$. Therefore, the part of the curve between $\theta = \pi$ and $\theta = 2\pi - \cos^{-1}(-1/e)$ is $A'N$; and for θ increasing from $2\pi - \cos^{-1}(-1/e)$ to 2π, r is positive and the points on the lower half of the branch LA are obtained. Here, the points L and N are on the curve at infinity.

Note. When $e = 0$, the equation of the conic reduces to $r = l$ which is clearly a circle of radius l with centre at the pole.

ILLUSTRATIVE EXAMPLES

Example 1. Prove that the curve given by

$$1/r = l\cos\theta + m\sin\theta + n$$

will, for non-zero values of l, m, n be a conic claiming the origin as a focus. Also prove that the curve takes a circular or straight line form according as one or other of the following sets of conditions is satisfied :

(a) $l = m = 0, n \neq 0$ (b) $n = 0$, l and $m \neq 0$.

Solution : The given curve is,

$$1/r = l \cos \theta + m \sin \theta + n \qquad \text{...(i)}$$

Putting $l = R \cos \alpha$ and $m = R \sin \alpha$ in (i), where R and α are arbitrary constants, we have

$$1/r = R \cos \alpha \cos \theta + R \sin \alpha \sin \theta + n$$

$$\Rightarrow \quad 1/r = R \cos (\theta - \alpha) + n$$

$$\Rightarrow \quad \frac{1/n}{r} = 1 + \frac{R}{n} \cos (\theta - \alpha)$$

$$\Rightarrow \quad \frac{L}{R} = 1 + E \cos (\theta - \alpha) \qquad \text{...(ii)}$$

where $L = \frac{1}{n}$ and $E = \frac{R}{n}$

This clearly represents a conic section whose focus is origin and the axis is inclined at an angle α with the initial line.

(a) When $l = m = 0, n \neq 0$, (i) reduces to

$$\frac{1}{r} = n, \quad \Rightarrow \quad r = \frac{1}{n} = \text{constant}$$

This represents a circle of radius $1/n$.

(b) When $n = 0$, l and $m \neq 0$, (i) becomes

$$\frac{1}{r} = l \cos \theta + m \sin \theta$$

$$\Rightarrow \quad 1 = lr \cos \theta + mr \sin \theta$$

$$\Rightarrow \quad 1 = lx + my$$

[changing into cartesian form]

This represents a straight line.

Example 2. Show that the equations

$$\frac{l}{r} = 1 - e \cos \theta \text{ and } \frac{l}{r} = -1 - e \cos \theta$$

represent the same conic.

Solution : The given equations are

$$\frac{l}{r} = 1 - e\cos\theta \quad \text{...(i)}$$

and

$$\frac{l}{r} = -1 - e\cos\theta \quad \text{...(ii)}$$

Let $P(r_1, \theta_1)$ be any point on (i), then we have

$$\frac{l}{r_1} = 1 - e\cos\theta_1 \quad \text{...(iii)}$$

The coordinates of P can also be taken as $(-r_1, \theta_1 + \pi)$ instead of (r_1, θ_1).

If it lies on (i), then

$$-\frac{l}{r_1} = 1 - e\cos(\theta_1 + \pi)$$

$$\Rightarrow \quad -\frac{l}{r_1} = 1 + e\cos\theta_1$$

$$\Rightarrow \quad \frac{l}{r_1} = -1 - e\cos\theta_1$$

$\therefore$ Locus of (r_1, θ_1) is

$$\frac{l}{r} = -1 - e\cos\theta$$

which is (ii).

But locus of (r_1, θ_1) was (i).

Hence, (i) and (ii) represent the same conic.

Example 3. In any conic, prove that the sum of the reciprocals of two perpendicular focal chords is constant.

Solution : Let the conic be

$$\frac{l}{r} = 1 + e\cos\theta \quad \text{...(i)}$$

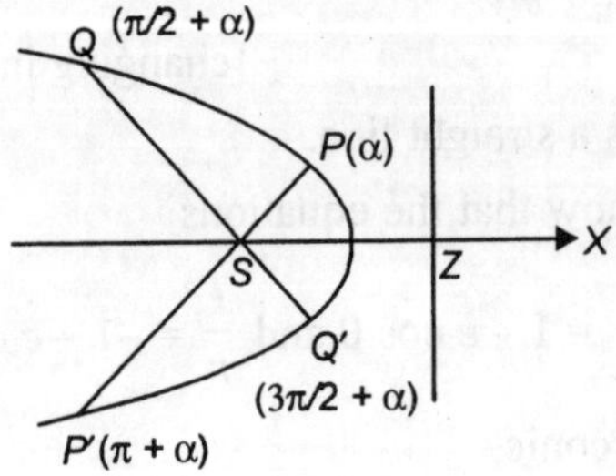

Fig.

Let PSP' and QSQ' be two mutually perpendicular focal chords. Let the vectorial angle of P be α. Then the vectorial angles of Q, P' and Q' will be $\pi/2 + \alpha$, $\pi + \alpha$ and $3\pi/2 + \alpha$ respectively.

Since, $P\ (SP, \alpha)$ lies in (i), we have

$$\frac{l}{SP} = 1 + e\cos\alpha$$

$$\Rightarrow \quad SP = \frac{l}{1 + e\cos\alpha} \qquad \text{...(ii)}$$

Since, $Q\left(SQ, \dfrac{\pi}{2} + \alpha\right)$ lies in (i), we have

$$\frac{l}{SQ} = 1 + e\cos\left(\frac{\pi}{2} + \alpha\right)$$

$$\Rightarrow \quad \frac{l}{SQ} = 1 - e\sin\alpha$$

$$\Rightarrow \quad SQ = \frac{l}{1 - e\sin\alpha} \qquad \text{...(iii)}$$

Since, $P'\ (SP', \pi + \alpha)$ lies on (i), we have

$$\frac{l}{SP'} = 1 + e\cos(\pi + \alpha)$$

$$\Rightarrow \quad \frac{l}{SP'} = 1 - e\cos\alpha$$

$$\Rightarrow \quad SP' = \frac{l}{1 - e\cos\alpha} \qquad \text{...(iv)}$$

Since, $Q'\left(SQ', \dfrac{3\pi}{2} + \alpha\right)$ lies on (i), we have

$$\frac{l}{SQ'} = 1 + e\cos\left(\frac{3\pi}{2} + \alpha\right)$$

$$\Rightarrow \quad \frac{l}{SQ'} = 1 + e\sin\alpha$$

$$\Rightarrow \quad SQ' = \frac{l}{1 + e\sin\alpha} \qquad \text{...(v)}$$

Now, $\quad PP' = SP + SP'$

$$= \frac{l}{1 + e\cos\alpha} + \frac{l}{1 - e\cos\alpha}$$

$$= \frac{2l}{1 - e^2\cos^2\alpha}$$

$$QQ' = SQ + SQ'$$

$$= \frac{l}{1 - e\sin\alpha} + \frac{l}{1 + e\sin\alpha}$$

$$= \frac{2l}{1 - e^2\sin^2\alpha}$$

$$\therefore \quad QQ' = \frac{1 - e^2\cos^2\alpha}{2l} + \frac{1 - e^2\sin^2\alpha}{2l}$$

$$= \frac{2 - e^2}{2l} = \text{constant.}$$

Example 4. Prove that the locus of the middle points of focal chords of a conic section is a conic section of the same kind.

Solution: Let the given conic be

$$\frac{l}{r} = 1 + e\cos\theta \qquad \ldots\text{(i)}$$

Let PSP' be any focal chord of this conic.

Let the vectorial angle of P be α. Then the vectorial angle of P' is $\pi + \alpha$.

$\because$ $P(SP, \alpha)$ lies on (i), we have

$$\frac{l}{SP} = 1 + e\cos\alpha$$

$$\Rightarrow \quad SP = \frac{l}{1 + e\cos\alpha} \qquad \ldots\text{(ii)}$$

$\because$ $P'(SP', \pi + \alpha)$ lies on (i), we have

$$\frac{l}{SP'} = 1 + e\cos(\pi + \alpha)$$

$$\Rightarrow \quad \frac{l}{SP'} = 1 - e\cos\alpha$$

$$\Rightarrow \quad SP' = \frac{l}{1 - e\cos\alpha} \qquad \ldots\text{(iii)}$$

Let M be the middle point of PSP'.

Let the coordinates of M be (R, ϕ). Then

$$\phi = \alpha$$

and

$$R = SM$$

$$= SP - PM$$

$$= SP - \frac{1}{2}PP'$$

$$= SP - \frac{1}{2}(SP + SP')$$

$$= \frac{SP - SP'}{2}$$

$$= \frac{1}{2}\left[\frac{l}{1 + e\cos\alpha} - \frac{l}{1 - e\cos\alpha}\right]$$

$$= -\frac{el\cos\alpha}{1 - e^2\cos^2\alpha}$$

$$= -\frac{el\cos\phi}{1 - e^2\cos^2\phi}.$$

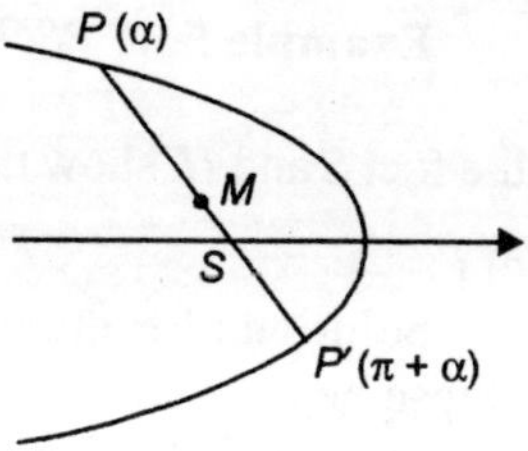

Fig.

Hence, the locus of the middle point $M(R, \phi)$ is

$$r = -\frac{el\cos\theta}{1 - e^2\cos^2\theta}$$

$$\Rightarrow \quad r(1 - e^2\cos^2\theta) + el\cos\theta = 0$$

$$\Rightarrow \quad r^2(1 - e^2\cos^2\theta) + elr\cos\theta = 0$$

$$\Rightarrow \quad r^2 - e^2r^2\cos^2\theta + elr\cos\theta = 0$$

$$\Rightarrow \quad x^2 + y^2 - e^2x^2 + el\,x = 0 \qquad \ldots\text{(iv)}$$

[changing into cartesian form]

Case I. If the original conic is a parabola, $e = 1$, then (iv) becomes,

$$x^2 + y^2 - x^2 + lx = 0$$

$$\Rightarrow \quad y^2 = -lx$$

which is also a parabola whose focus is S and latus rectum l.

Case II. If $e \neq 1$, then equation (iv) becomes,

$$(1 - e^2)x^2 + y^2 + lex = 0$$

$$\Rightarrow \quad (1 - e^2)\left\{x^2 + \frac{elx}{1 - e^2}\right\} + y^2 = 0$$

$$\Rightarrow \quad (1 - e^2)\left\{x + \frac{el}{2(1 - e^2)}\right\}^2 + y^2 = \frac{e^2l^2}{4(1 - e^2)}$$

$$\Rightarrow \quad \left\{x + \frac{el}{2(1 - e^2)}\right\}^2 + \frac{y^2}{1 - e^2} = \left\{\frac{el}{2(1 - e^2)}\right\}^2$$

which represents an ellipse or hyperbola according as $e < 1$ or $e > 1$, i.e., according as the original conic is an ellipse or hyperbola.

Example 5. If *PSQ*, *PHR* be two chords of an ellipse through the foci *S* and *H*, show that $\frac{PS}{SQ}+\frac{PH}{HR}$ is independent of the position of *P*.

Solution : Let the equation of the ellipse be,

$$\frac{l}{r} = 1 + e\cos\theta \qquad \text{...(i)}$$

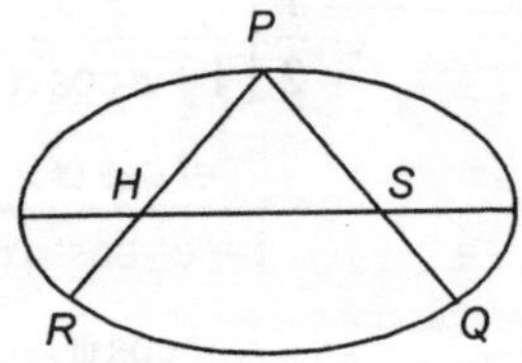

Fig.

Let the vectorial angle of *P* be α. Then the vectorial angle of *Q* is π + α.

Since, *P* (*SP*, α) lies on (i), we have

$$\frac{l}{SP} = 1 + e\cos\alpha$$

$$\Rightarrow \quad \frac{1}{SP} = \frac{1+e\cos\alpha}{l} \qquad \text{...(ii)}$$

Since, *Q* {*SQ* (π + α)} lies on (i), we have

$$\Rightarrow \quad \frac{l}{SQ} = 1 + e\cos(\pi+\alpha) = 1 - e\cos\alpha$$

$$\Rightarrow \quad \frac{1}{SQ} = \frac{1-e\cos\alpha}{l} \qquad \text{...(iii)}$$

$$\therefore \quad \frac{1}{SP}+\frac{1}{SQ} = \frac{1+e\cos\alpha}{l}+\frac{1-e\cos\alpha}{l} = \frac{2}{l}$$

$$\Rightarrow \quad 1+\frac{SP}{SQ} = \frac{2}{l}SP$$

$$\Rightarrow \quad \frac{SP}{SQ} = \frac{2}{l}SP - 1 \qquad \text{...(iv)}$$

Similarly, $\frac{HP}{HR} = \frac{2}{l}HP - 1$ ...(v)

Adding (iv) and (v), we get

$$\frac{SP}{SQ}+\frac{HP}{HR} = \frac{2}{l}(SP+HP) - 2$$

$$= \frac{2}{l}(\text{Major axis of the ellipse}) - 2$$

$$= \frac{2}{l}(2a) - 2 = \frac{4a}{l} - 2$$

which being free from α, is independent of the position of *P*.

Example 6. A circle passes through the focus S of a conic and meets it in four points whose distances from S are r_1, r_2, r_3 and r_4. Prove that

(1) $r_1r_2r_3r_4 = \dfrac{d^2l^2}{e^2}$ (2) $\dfrac{1}{r_1}+\dfrac{1}{r_2}+\dfrac{1}{r_3}+\dfrac{1}{r_4}=\dfrac{2}{l}$

where $2l$ and e are the latus rectum and eccentricity of the conic, and d is the diameter of the circle.

Solution : Let us take the focus S as the pole and the axis of the conic as the initial line. Then the equation of the conic can be taken as

$$\frac{l}{r} = 1 + e\cos\theta \qquad \text{...(i)}$$

Let the diameter of the circle which passes through S be inclined at an angle γ to the axis, then the equation of the circle can be taken as

$$r = d\cos(\theta - \gamma)$$

$$\Rightarrow \quad r = d(\cos\theta\cos\gamma + \sin\theta\sin\gamma) \qquad \text{...(ii)}$$

From (i), $\cos\theta = \dfrac{1}{e}\left(\dfrac{l}{r}-1\right) = \dfrac{l-r}{er}$

$$\therefore \quad \sin\theta = \sqrt{1-\cos^2\theta} = \sqrt{1-\left(\frac{l-r}{er}\right)^2}$$

Substituting the values of $\cos\theta$ and $\sin\theta$ in (ii), we get

$$r = d\left[\frac{l-r}{er}\cos\gamma + \sqrt{1-\left(\frac{1-r}{er}\right)^2}\sin\gamma\right]$$

$$\Rightarrow [er^2 - d(1-r)\cos\gamma]^2 - d^2[e^2r^2 - (l-r)^2]\sin^2\gamma$$

$$\Rightarrow e^2r^4 + 2de\cos\gamma\, r^3 + (a^2 - 2eld\cos\gamma - e^2d^2\sin^2\gamma)\, r^2 - 2ld^2r + d^2l^2 = 0 \qquad \text{...(iii)}$$

Let r_1, r_2, r_3, r_4 be the roots of equation (iii). Then,

$$r_1r_2r_3r_4 = \frac{d^2l^2}{e^2} \qquad \text{...(iv)}$$

and $$r_2r_3r_4 + r_1r_3r_4 + r_1r_2r_4 + r_1r_2r_3 = \frac{2ld^2}{e^2} \qquad \text{...(v)}$$

Dividing (v) by (iv), we get

$$\frac{1}{r_1}+\frac{1}{r_2}+\frac{1}{r_3}+\frac{1}{r_4}=\frac{2}{l}.$$

Example 7. A straight line drawn through the common focus S of a number of conics meets them in the points P_1, P_2 .... On it is taken a point Q such that the reciprocal of SQ is equal to the sum of the reciprocals of SP_1, SP_2.... Prove that the locus of Q is a conic section whose focus is S and the reciprocal of whose latus rectum is equal to the sum of the reciprocals of the latera recta of the given conics.

Solution : Take the common focus S as pole. Let SX be the initial line.

$$\frac{l_1}{r} = 1 + e_1 \cos(\theta - \alpha_1)$$

$$\frac{l_2}{r} = 1 + e_2 \cos(\theta - \alpha_2), \;.....$$

Let the given line through S makes an angle ξ with SX.

Then
$$\frac{l_1}{SP_1} = 1 + e_1 \cos(\xi - \alpha_1)$$

$$\frac{l_2}{SP_2} = 1 + e_2 \cos(\xi - \alpha_2), \;..... \text{ etc.}$$

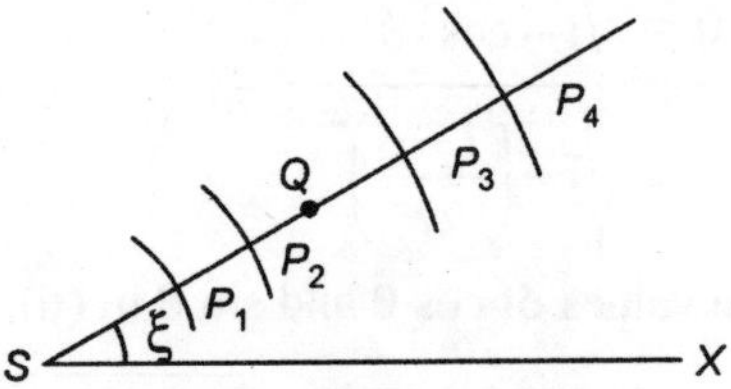

Fig.

Since, Q is a point on this line, such that

$$\frac{1}{SQ} = \frac{1}{SP_1} + \frac{1}{SP_2} + ...\text{(given)}$$

$$\therefore \quad \frac{1}{SQ} = \frac{1 + e_1 \cos(\xi - \alpha_1)}{l_1} + \frac{1 + e_2 \cos(\xi - \alpha_2)}{l_2} + ...$$

$$= \left(\frac{1}{l_1} + \frac{1}{l_2} + ...\right) + \cos\xi \left\{\frac{e_1}{l_1}\cos\alpha_1 + \frac{e_2}{l_2}\cos\alpha_2 + ...\right\}$$

$$+ \sin\xi \left\{\frac{e_1}{l_1}\sin\alpha_1 + \frac{e_2}{l_2}\sin\alpha_2 + ...\right\}$$

$$= \frac{1}{L} + \cos\xi \frac{E}{L}\cos\xi + \sin\xi \frac{E}{L}\sin\alpha$$

where $$\frac{1}{L} = \frac{1}{l_1} + \frac{1}{l_2} + ...$$

$$\Rightarrow \quad \frac{1}{2L} = \frac{1}{2l_1} + \frac{1}{2l_2} + ...$$

$$\frac{E}{L}\cos\alpha = \frac{e_1}{l_1}\cos\alpha_1 + \frac{e_2}{l_2}\cos\alpha_2 + ...$$

and $$\frac{E}{L}\sin\alpha = \frac{e_1}{l_1}\sin\alpha_1 + \frac{e_2}{l_2}\sin\alpha_2 + ...$$

$$\therefore \quad \frac{1}{SQ} = \frac{1}{L} + \frac{E}{L}\cos(\xi - \alpha)$$

Hence, locus of Q is,

$$\frac{1}{r} = \frac{1}{L} + \frac{E}{L}\cos(\theta - \alpha) \Rightarrow \frac{L}{r} = 1 + E\cos(\theta - \alpha)$$

which is a conic whose focus is S and latus rectum $2L$ is given by

$$\frac{1}{2L} = \frac{1}{2l_1} + \frac{1}{2l_2} + ...$$

EXERCISE 2 (A)

1. Show that in a conic the semi latus-rectum is the harmonic mean between the segments of a focal chord.

2. (i) Show that the equations $l/r = 1 + e\cos\theta$ and $l/r = -1 + e\cos\theta$ represent the same conic.

(ii) Prove that $l/r = 1 - 3\cos\theta$, and $l/r = -1 - 3\cos\theta$ represent same conic.

3. In any conic, prove that the sum of the reciprocals of the segments cut of by the focus of any focal chord is constant.

or

If PSP' is the chord of a conic then prove that $\frac{1}{SP} + \frac{1}{SP'} = \frac{2}{l}$ where l is the semi-latus rectum.

4. In a parabola, prove that the length of a focal chord which is inclined at 30° to the axis is four times the length of the latus rectum.

5. Prove that the perpendicular focal chords of a rectangular hyperbola are equal.

6. If PSP' and QSQ' are two perpendicular focal chords of a conic,

prove that $\dfrac{1}{SP \cdot SP'} + \dfrac{1}{SQ \cdot SQ'}$ = constant.

7. A point P moves so that the sum of its distances from two fixed points S, S' is constant and equal to $2a$. Show that P lies on the conic, $\dfrac{a(1-e^2)}{r} = 1 - e \cos \theta$ referred to S as pole and SS' as initial line, SS' being equal to $2ae$.

8. In the ellipse, $2/r = 1 - e \cos \theta$, if PQ be any chord through the focus S which has been taken as pole, prove that $\dfrac{1}{SP} + \dfrac{1}{SQ} = 1$.

9. If two conics have a common focus, show that the two of their common chords will pass through the point of intersection of their directrices.

10. A circle of given radius passing through the focus S of a given conic intersects it in A, B, C, D. Show that $SA \cdot SB \cdot SC \cdot SD$ is constant.

11. A variable circle always passes through a fixed point S, and cuts a conic in the points A, B, C, D. Show that the continued product of the distances of the points of intersection from the fixed point varies as the square of the radius of the circle.

12. If SP, a focal chord of a conic, is produced to Q, so that

$$SQ = k \cdot SP$$

where k is a constant, prove that the locus of Q is a conic of the same eccentricity as of the given conic, and the latus rectum is k times that of the given conic.

2.5 Chord Joining Two Points

To find the equation of the chord joining two points on a conic.

Let the equation of the conic be

$$\frac{l}{r} = 1 + e \cos \theta \quad \ldots(1)$$

Let α, β be the vectorial angles of two points P and Q respectively on (1).

Let the equation of chord PQ be,

$$\frac{l}{r} = A \cos \theta + B \sin \theta \quad \ldots(2)$$

where A and B are the constants to be determined.

$\because P(SP, \alpha)$, lies on (1) and (2), we have

$$\frac{l}{SP} = 1 + e\cos\alpha$$

and $$\frac{l}{SP} = A\cos\alpha + B\sin\alpha.$$

Equating the two values of l/SP, we get

$$1 + e\cos\alpha = A\cos\alpha + B\sin\alpha$$

$$\Rightarrow (A - e)\cos\alpha + B\sin\alpha - 1 = 0 \qquad \ldots(3)$$

$\because Q(SQ, \beta)$, lies on (1) and (2), we have

$$\therefore \qquad \frac{l}{SQ} = 1 + e\cos\beta$$

and $$\frac{l}{SQ} = A\cos\beta + B\sin\beta$$

Equating the two values of l/SQ, we get

$$1 + e\cos\beta = A\cos\beta + B\sin\beta$$

$$\Rightarrow (A - e)\cos\beta + B\sin\beta - 1 = 0 \qquad \ldots(4)$$

Solving (3) and (4) for $(A - e)$ and B, we get

$$\frac{A-e}{-\sin\alpha + \sin\beta} = \frac{\beta}{-\cos\beta + \cos\alpha} = \frac{1}{\cos\alpha\,\sin\beta - \cos\beta\,\sin\alpha}$$

$$\Rightarrow \qquad \frac{A-e}{\sin\beta - \sin\alpha} = \frac{\beta}{\cos\alpha - \cos\beta} = \frac{l}{\sin(\beta - \alpha)}$$

$$\Rightarrow \qquad \frac{A-e}{2\cos\frac{\beta+\alpha}{2}\sin\frac{\beta-\alpha}{2}} = \frac{B}{2\sin\frac{\alpha+\beta}{2}\sin\frac{\beta-\alpha}{2}}$$

$$= \frac{1}{2\sin\left(\frac{\beta-\alpha}{2}\right)\cos\left(\frac{\beta-\alpha}{2}\right)}$$

$$\Rightarrow \qquad \frac{A-e}{\cos\left(\frac{\beta+\alpha}{2}\right)} = \frac{B}{\sin\left(\frac{\alpha+\beta}{2}\right)} = \frac{1}{\cos\left(\frac{\beta-\alpha}{2}\right)}$$

$$\Rightarrow \qquad A = e + \sec\left(\frac{\beta-\alpha}{2}\right)\cos\left(\frac{\alpha+\beta}{2}\right)$$

$$B = \sec\left(\frac{\beta-\alpha}{2}\right)\sin\left(\frac{\alpha+\beta}{2}\right)$$

Substituting the values of A and B in (2), the equation of chord PQ is,

$$\frac{l}{r} = \left[e + \sec\left(\frac{\beta-\alpha}{2}\right)\cos\left(\frac{\alpha+\beta}{2}\right)\right]\cos\theta + \sec\left(\frac{\beta-\alpha}{2}\right)\sin\left(\frac{\alpha+\beta}{2}\right)\sin\theta$$

$$\Rightarrow \quad \frac{l}{r} = e\cos\theta + \sec\left(\frac{\beta-\alpha}{2}\right)\left(\cos\theta\cos\frac{\alpha+\beta}{2} + \sin\theta\sin\frac{\alpha+\beta}{2}\right)$$

$$\Rightarrow \quad \frac{l}{r} = e\cos\theta + \sec\left(\frac{\beta-\alpha}{2}\right)\cos\left(\theta - \frac{\alpha+\beta}{2}\right) \qquad \ldots(5)$$

which is the required equation of chord PQ.

Note 1. The equation of the chord joining two points $\alpha - \beta$ and $\alpha + \beta$ on the conic $l/r = 1 + e\cos\theta$, is given by,

$$\frac{l}{r} = e\cos\theta + \sec\beta\cos(\theta - \alpha)$$

Note 2. The equation of the chord joining the points α and β on the conic $l/r = 1 + e\cos(\theta - \gamma)$, is given by,

$$\frac{l}{r} = e\cos(\theta - \gamma) + \sec\left(\frac{\beta-\alpha}{2}\right)\cos\left\{\theta - \frac{\alpha+\beta}{2}\right\}.$$

Note 3. The equation of the chord joining the points $\alpha - \beta$ and $\alpha + \beta$, on the conic,

$$\frac{l}{r} = 1 + e\cos(\theta - \gamma),$$

is given by,

$$\frac{l}{r} = e\cos(\theta - \gamma) + \sec\beta\cos(\theta - \alpha).$$

Note 4. If the equation of the conic is,

$$\frac{l}{r} = 1 - e\cos\theta \text{ or } \frac{l}{r} = 1 - e\cos(\theta - \gamma),$$

replace e by $-e$ in the corresponding cases.

2.6 Equation of Tangent

To find the equation of the tangent at a point 'α' of the conic,

$$\frac{l}{r} = 1 + e \cos \theta.$$

By Art. 2.5, we know that the equation of the chord joining the point P ('α') and Q ('β') on the conic $l/r = 1 + e \cos \theta$, is given by

$$\frac{l}{r} = e \cos \theta + \sec\left(\frac{\beta - \alpha}{2}\right)\cos\left\{\theta - \frac{\alpha + \beta}{2}\right\} \quad \text{...(1)}$$

when $Q \rightarrow P$, i.e., when $\beta \rightarrow \alpha$, (1) becomes tangent at P ('α') to the conic, $l/r = 1 + e \cos \theta$. Hence, the equation of the tangent is,

$$\frac{l}{r} = e \cos \theta + \cos (\theta - \alpha).$$

Note 1. The equation of the tangent at the point 'α' of the conic, $l/r = 1 - e \cos \theta$ is,

$$\frac{l}{r} = - e \cos \theta + \cos (\theta - \alpha).$$

Note 2. The equation of the tangent at the point 'α' of the conic, $l/r = 1 + e \cos (\theta - \gamma)$ is,

$$\frac{l}{r} = e \cos (\theta - \gamma) + \cos (\theta - \alpha).$$

Note 3. The equation of the tangent at the point 'α' of the conic, $l/r = 1 - e \cos (\theta - \gamma)$ is,

$$\frac{l}{r} = - e \cos (\theta - \gamma) + \cos (\theta - \alpha).$$

2.7 Asymptotes

To find the equation of the asymptotes of the conic,

$$l/r = 1 + e \cos \theta$$

The conic is,

$$\frac{l}{r} = 1 + e \cos \theta \quad \text{...(1)}$$

Let (r_1, α) be a point on (1). Then,

$$\frac{l}{r_1} = 1 + e \cos \alpha \quad \text{...(2)}$$

The equation of the tangent of (1) at the point 'α' is,

$$\frac{l}{r} = e \cos \theta + \cos (\theta - \alpha)$$

$$\Rightarrow \qquad \frac{l}{r} = (e + \cos\alpha)\cos\theta + \sin\alpha\sin\theta \qquad \ldots(3)$$

This tangent will tend to an asymptote, when the point of contact (r_1, α) tends to infinity, i.e., $r_1 \to \infty$.

Therefore, putting $r_1 = \infty$ in (2), we have

$$1 + e\cos\alpha = 0$$

$$\Rightarrow \qquad \cos\alpha = -\frac{1}{e}$$

$$\therefore \qquad \sin\alpha = \pm\sqrt{1 - \frac{1}{e^2}}$$

$\therefore$ (3) can be written as,

$$\frac{l}{r} = \left(e - \frac{1}{e}\right)\cos\theta \pm \sqrt{1 - \frac{1}{e^2}}\sin\theta$$

$$\Rightarrow \qquad \frac{el}{r} = (e^2 - 1)\cos\theta \pm \sqrt{e^2 - 1}\sin\theta$$

which is the required equation of the two asymptotes.

These asymptotes will be real, if $e^2 - 1 > 0$

i.e. $\quad e^2 > 1$

i.e. $\quad e > 1$

i.e., the conic is a hyperbola.

ILLUSTRATIVE EXAMPLES

Example 1. Find the condition that the straight line

$$l/r = A\cos\theta + B\sin\theta$$

may be a tangent to the conic,

$$l/r = 1 + e\cos(\theta - \theta')$$

Solution : Let the straight line

$$l/r = A\cos\theta + B\sin\theta \qquad \ldots(i)$$

touches the conic

$$l/r = 1 + e\cos(\theta - \theta') \qquad \ldots(ii)$$

at 'α'.

Then the equation of the tangent to (ii) at 'α' is,

$$\frac{l}{r} = e\cos(\theta - \theta') + \cos(\theta - \alpha)$$

$$\Rightarrow \frac{l}{r} = e(\cos\theta\cos\theta' + \sin\theta\sin\theta') + (\cos\theta\cos\alpha + \sin\theta\sin\alpha)$$

$$\Rightarrow \frac{l}{r} = (e\cos\theta' + \cos\alpha)\cos\theta + (e\sin\theta' + \sin\alpha)\sin\theta \quad \text{...(iii)}$$

Lines (i) and (iii) must be identical.

Hence, comparing the coefficients of (i) and (iii), we have

$$\frac{e\cos\theta' + \cos\alpha}{A} = \frac{e\sin\theta' + \sin\alpha}{B} = 1$$

$$\Rightarrow \qquad \cos\alpha = A - e\cos\theta'$$

and $\qquad \sin\alpha = B - e\sin\theta'.$

On squaring and adding, we have

$(A - e\cos\theta')^2 + (B - e\sin\theta')^2 = \cos^2\alpha + \sin^2\alpha = 1$

which is the required condition.

Example 2. *PSP′* is a focal chord of a conic. Prove that the angle between the tangents at *P* and *P′* is,

$$\tan^{-1}\left(\frac{2e\sin\alpha}{1-e^2}\right)$$

where α is the angle between the chord and the axis.

Solution : Let the conic be

$$l/r = 1 + e\cos\theta \qquad \text{...(i)}$$

Since, *PSP′* is a focal chord of (i), therefore, if the vectorial angle of *P* is α, then the vectorial angle of *P′* is π + α.

The equation of the tangent at *P* (α) is,

$$l/r = e\cos\theta + \cos(\theta - \alpha)$$

$$\Rightarrow \qquad l/r = (e + \cos\alpha)\cos\theta + \sin\alpha\sin\theta$$

$$\Rightarrow \qquad l = (e + \cos\alpha)\, r\cos\theta + \sin\alpha \cdot r\sin\theta$$

$$\Rightarrow \qquad l = (e + \cos\alpha)\, x + \sin\alpha \cdot y.$$

(changing into cartesian form)

∴ The slope of the tangent at *P* (α) is,

$$-\frac{e + \cos\alpha}{\sin\alpha} = m_1 \text{ (say).}$$

Similarly, the slope of the tangent at $P'(\pi + \alpha)$ is,

$$= -\frac{e + \cos(\pi + \alpha)}{\sin(\pi + \alpha)} \quad \text{(replacing } \alpha \text{ by } \pi + \alpha\text{)}$$

$$= \frac{e - \cos\alpha}{\sin\alpha} = m_2 \text{ (say).}$$

Let ϕ be the angle between the tangents at P and P', then

$$\tan\phi = \frac{m_1 \sim m_2}{1 + m_1 m_2}$$

$$= \frac{\frac{e - \cos\alpha}{\sin\alpha} + \frac{e + \cos\alpha}{\sin\alpha}}{1 - \frac{e - \cos\alpha}{\sin\alpha} \cdot \frac{e + \cos\alpha}{\sin\alpha}}$$

$$= \frac{2e\sin\alpha}{\sin^2\alpha - (e^2 - \cos^2\alpha)}$$

$$= \frac{2e\sin\alpha}{1 - e^2}$$

$$\therefore \qquad \phi = \tan^{-1}\left(\frac{2e\sin\alpha}{1 - e^2}\right).$$

Example 3. Prove that the two conics

$\frac{l_1}{r} = 1 + e_1 \cos\theta$ and $\frac{l_2}{r} = 1 + e_2 \cos(\theta - \alpha)$

will touch one another if

$l_1^2(1 - e_2^2) + l_2^2(1 - e_1^2) = 2l_1 l_2(1 - e_1 e_2 \cos\alpha)$.

Solution : The given conics are

$$\frac{l_1}{r} = 1 + e_1 \cos\theta \qquad \text{...(i)}$$

and $$\frac{l_2}{r} = 1 + e_2 \cos(\theta - \alpha) \qquad \text{...(ii)}$$

Let the vectorial angle of the point of contact of the conics (i) and (ii) be β. Then, the equation of the tangents at the point β to the two conics are,

$$\frac{l_1}{r} = e_1 \cos\theta + \cos(\theta - \beta)$$

$$= (e_1 + \cos\beta)\cos\theta + \sin\beta\sin\theta \qquad \text{...(iii)}$$

and $$\frac{l_2}{r} = e_2 \cos(\theta - \alpha) + \cos(\theta - \beta)$$

$$= (e_2 \cos\alpha + \cos\beta)\cos\theta + (e_2 \sin\alpha + \sin\beta)\sin\theta \quad ...(iv)$$

Since, the conics (i) and (ii) touch each other at 'β', hence, equations (iii) and (iv) are identical.

Therefore, comparing the coefficients, we get

$$\frac{e_1 + \cos\beta}{e_2 \cos\alpha + \cos\beta} = \frac{\sin\beta}{e_2 \sin\alpha + \sin\beta} = \frac{l_1}{l_2}$$

$$\Rightarrow \quad \sin\beta = -\frac{l_1 e_2 \sin\alpha}{l_1 - l_2} \quad ...(v)$$

and $$\cos\beta = -\frac{l_1 e_2 \cos\alpha + l_2 e_1}{l_1 - l_2} \quad ...(vi)$$

Squaring and adding (v) and (vi), we get

$$\sin^2\beta + \cos^2\beta = 1 = \frac{l_2^2 e_2^2 \sin^2\alpha}{(l_1 - l_2)^2} + \frac{(l_2 e_1 - l_1 e_2 \cos\alpha)^2}{(l_1 - l_2)^2}$$

$$\Rightarrow \quad (l_1 - l_2)^2 = l_1^2 e_2^2 \sin^2\alpha + (l_2 e_1 - l_1 e_2 \cos\alpha)^2$$

$$\Rightarrow l_1^2(1 - e_2^2) + l_2^2(1 - e_1^2) = 2l_1 l_2 (1 - e_1 e_2 \cos\alpha).$$

Example 4. Find the condition that conic

$$\frac{l_1}{r} = 1 + e_1 \cos\theta \text{ and } \frac{l_2}{r} = 1 + e_2 \cos\theta$$

will touch each other.

Solution : Putting $\alpha = 0$ in Ex. 3, we get required condition.

Example 5. A conic is described having the same focus and eccentricity as the conic $l/r = 1 + e\cos\theta$ and the two conics touch at the point $\theta = \alpha$; prove that the length of its latus rectum is,

$$\frac{2l(1 - e^2)}{e^2 + 2e\cos\alpha + 1}$$

and further prove that the angle between their axes is,

$$2\tan^{-1}\left(-\frac{e + \cos\alpha}{\sin\alpha}\right).$$

Solution : The given conic is,

$$\frac{l}{r} = 1 + e\cos\theta \quad ...(i)$$

Equation of the conic having the same focus and eccentricity as (i) can be taken as

$$\frac{L}{r} = 1 + e\cos(\theta - \gamma) \qquad \text{...(ii)}$$

so that γ is the angle between their axes.

Equation of the tangent at the point α to the conic (i) is,

$$l/r = e\cos\theta + \cos(\theta - \alpha)$$

$$\Rightarrow \quad l/r = (e + \cos\alpha)\cos\theta + \sin\alpha\sin\theta \qquad \text{...(iii)}$$

Equation of the tangent at the point α to the conic (ii) is,

$$L/r = e\cos(\theta - \gamma) + \cos(\theta - \alpha)$$

$$\Rightarrow \quad L/r = (e\cos\gamma + \cos\alpha)\cos\theta + (e\sin\gamma + \sin\alpha)\sin\theta \qquad \text{...(iv)}$$

Since, the conics (i) and (ii) touch at $\theta = \alpha$, hence, the lines (iii) and (iv) must be identical.

Therefore, comparing the coefficients in (iii) and (iv), we get

$$\frac{L}{l} = \frac{e\cos\gamma + \cos\alpha}{e + \cos\alpha} = \frac{e\sin\gamma + \sin\alpha}{\sin\alpha}$$

$$\Rightarrow \frac{L}{l}\{(e + \cos\alpha) - \cos\alpha = e\cos\gamma \qquad \text{...(v)}$$

and

$$\frac{L}{l}\sin\alpha - \sin\alpha = e\sin\gamma \qquad \text{...(vi)}$$

Squaring and adding (v) and (vi), we get

$$\frac{L^2}{l^2}(e + \cos\alpha)^2 + \sin^2\alpha\} - \frac{2L}{l}\{(e + \cos\alpha)\cos\alpha + \sin^2\alpha\}$$
$$+ \cos^2\alpha + \sin^2\alpha = e^2(\cos^2\gamma + \sin^2\gamma)$$

$$\Rightarrow (e^2 + 2e\cos\alpha + 1)\frac{L^2}{l^2} - \frac{2L}{l}(e\cos\alpha + 1) + 1 - e^2 = 0$$

$$\Rightarrow (e^2 + 2e\cos\alpha + 1)x^2 - 2x(e\cos\alpha + 1) + 1 - e^2 = 0 \text{ if } L/l = x$$

$$\Rightarrow (x - 1)[(e^2 + 2e\cos\alpha + 1)x - (1 - e^2)] = 0$$

$$\Rightarrow \quad x = 1 \text{ or } \frac{1 - e^2}{e^2 + 2e\cos\alpha + 1}$$

If $x = 1$, then $L = 1$ which is inadmissible.

If $x = \dfrac{1 - e^2}{e^2 + 2e\cos\alpha + 1}$, then

$$\frac{L}{l} = \frac{1 - e^2}{e^2 + 2e\cos\alpha + 1} \qquad \text{...(vii)}$$

$$\therefore \quad L = \frac{(1 - e^2)l}{e^2 + 2e\cos\alpha + 1}$$

$\therefore$ Length of the latus rectum of (ii)

$$= 2L$$

$$= \frac{2l(1-e^2)}{e^2+2e\cos\alpha+1}.$$

Substituting the value of L/l from equation (vii) in (vi), we get

$$\frac{(1-e^2)\sin\alpha}{e^2+2e\cos\alpha+1} - \sin\alpha = e\sin\gamma$$

$$\Rightarrow \qquad \sin\gamma = \frac{\sin\alpha}{e}\left[\frac{1-e^2}{e^2+2e\cos\alpha+1}-1\right]$$

$$= \frac{\sin\alpha}{e}\left[\frac{1-e^2-e^2-2e\cos\alpha-1}{e^2+2e\cos\alpha+1}\right]$$

$$= \frac{\sin\alpha}{e}\left[\frac{-2e(e+\cos\alpha)}{e^2+2e\cos\alpha+1}\right]$$

$$= \frac{-2\sin\alpha(e+\cos\alpha)}{e^2+2e\cos\alpha+1}$$

Applying componendo and dividendo, we get

$$\frac{1-\sin\gamma}{1+\sin\gamma} = \frac{e^2+2e\cos\alpha+1+2\sin\alpha(e+\cos\alpha)}{e^2+2e\cos\alpha+1-2\sin\alpha(e+\cos\alpha)}$$

$$\Rightarrow \frac{(\sin\gamma/2-\cos\gamma/2)^2}{(\sin\gamma/2+\cos\gamma/2)^2} = \frac{(e+\cos\alpha+\sin\alpha)^2}{(e+\cos\alpha-\sin\alpha)^2}$$

Taking square root and dividing the numerator and denominator of the left hand side by $\cos\gamma/2$,

$$\frac{\tan\gamma/2-1}{\tan\gamma/2+1} = \frac{e+\cos\alpha+\sin\alpha}{e+\cos\alpha-\sin\alpha}.$$

Again, by componendo and dividendo,

$$\frac{2\tan\gamma/2}{2} = \frac{2(e+\cos\alpha)}{-2\sin\alpha}$$

$$\Rightarrow \qquad \tan\frac{\gamma}{2} = -\left(\frac{e+\cos\alpha}{\sin\alpha}\right)$$

$$\Rightarrow \qquad \gamma = 2\tan^{-1}\left(-\frac{e+\cos\alpha}{\sin\alpha}\right).$$

Example 6. If the tangents at any two points P and Q of a conic meet in a point T, and if the straight line PQ meets the directrix corresponding to S in a point K, then the angle KST is a right angle.

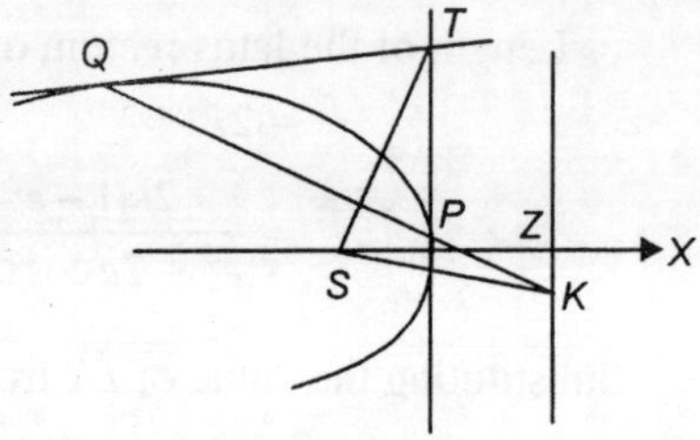

Fig.

Solution : Let the equation of the conic be

$$\frac{l}{r} = 1 + e\cos\theta \qquad \ldots\text{(i)}$$

Let vectorial angles of P and Q be α and β respectively. Then the equation of the chord PQ is,

$$\frac{l}{r} = e\cos\theta + \sec\left(\frac{\beta-\alpha}{2}\right)\cos\left(\theta - \frac{\alpha+\beta}{2}\right) \qquad \ldots\text{(ii)}$$

Also, the equation of the directrix corresponding to S is,

$$l/r = e\cos\theta \qquad \ldots\text{(iii)}$$

Equations (ii) and (iii) intersect at K.

$\therefore$ At K,

$$e\cos\theta = \cos\theta + \sec\left(\frac{\beta-\alpha}{2}\right)\cos\left(\theta - \frac{\alpha+\beta}{2}\right)$$

$$\Rightarrow \sec\left(\frac{\beta-\alpha}{2}\right)\cos\left(\theta - \frac{\alpha+\beta}{2}\right) = 0$$

Since, $\quad \sec\left(\frac{\beta-\alpha}{2}\right) \neq 0$

$$\therefore \quad \cos\left(\theta - \frac{\alpha+\beta}{2}\right) = 0 = \cos\frac{\pi}{2}$$

$$\therefore \quad \theta = \frac{\pi}{2} + \frac{\alpha+\beta}{2}$$

$\therefore$ Vectorial angle of $\quad K = \dfrac{\pi}{2} + \dfrac{\alpha+\beta}{2}$.

Now, equation of tangent at P is,

$$\frac{l}{r} = e\cos\theta + \cos(\theta - \alpha) \qquad \ldots\text{(iv)}$$

and equation of tangent at Q is,

$$\frac{l}{r} = e\cos\theta + \cos(\theta - \beta) \qquad \ldots\text{(v)}$$

Equations (iv) and (v) intersect at T.

$\therefore$ At T

$$e\cos\theta + \cos(\theta-\alpha) = e\cos\theta + \cos(\theta-\beta)$$

$$\Rightarrow \quad \cos(\theta-\alpha) = \cos(\theta-\beta).$$

Since, $\alpha \neq \beta$

$$\therefore \quad \cos(\theta-\alpha) = \cos(\beta-\theta) \qquad [\because \cos(-\theta) = \cos\theta]$$

$$\Rightarrow \quad \theta-\alpha = \beta-\theta$$

$$\Rightarrow \quad 2\theta = \alpha+\beta$$

$$\Rightarrow \quad \theta = \frac{\alpha+\beta}{2}$$

$\therefore$ Vectorial angle of $T = \dfrac{\alpha+\beta}{2}$

Difference of the vectorial angles of K and T

$$= \left(\frac{\pi}{2}+\frac{\alpha+\beta}{2}\right) - \left(\frac{\alpha+\beta}{2}\right) = \frac{\pi}{2}$$

Hence, $\angle KST = \dfrac{\pi}{2}$.

Example 7. If T be any point on the tangent at P of a conic, TR the perpendicular on the radius vector SP, S being the focus of the conic, TD the perpendicular on the directrix, then show that

$$SR = e \cdot TD.$$

Solution : Let the equation of the conic be

$$\frac{l}{r} = 1 + e\cos\theta \qquad \ldots\text{(i)}$$

Let the vectorial angle of P be α.

Then, the equation of the tangent at P is,

$$\frac{l}{r} = e\cos\theta + \cos(\theta-\alpha) \qquad \ldots\text{(ii)}$$

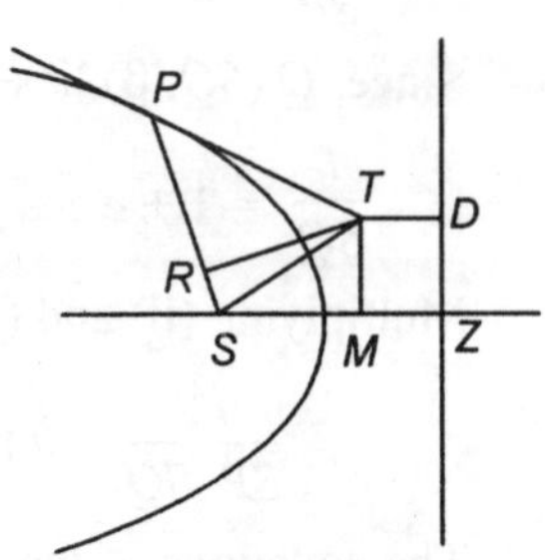

Fig.

Let the vectorial angle of T be β.

Then, $T(ST, \beta)$ lies on (ii).

$$\therefore \quad \frac{l}{ST} = e\cos\beta + \cos(\beta-\alpha)$$

$$\Rightarrow \quad l = e \cdot ST\cos\beta + ST\cos(\beta-\alpha)$$

$$= e \cdot ST \cos \beta + ST \cos (\alpha - \beta)$$
$$= e \cdot ST \cos \beta + ST \cos (\angle TSR)$$
$$= e \cdot ST \cos \beta + SR$$

$$\therefore \qquad SR = l - e \cdot ST \cos \beta$$
$$= l - e \cdot SM$$
$$= l - e\,(SZ - MZ)$$
$$= l - e\,(SZ - TD)$$
$$= l - e\left(\frac{l}{e} - TD\right) = l - l + e \cdot TD = e \cdot TD.$$

Example 8. If the tangents at P and Q of a conic meet in a point T and S be the focus; prove that if the conic be central

$$\frac{1}{SP \cdot SQ} - \frac{1}{ST^2} = \frac{1}{b^2} \sin^2 \frac{PSQ}{2}$$

where b is the semi-minor axis of the conic.

Solution : Let the equation of the conic be,

$$\frac{l}{r} = 1 + e \cos \theta \qquad \text{...(i)}$$

Let the vectorial angles of P and Q be α and β respectively.

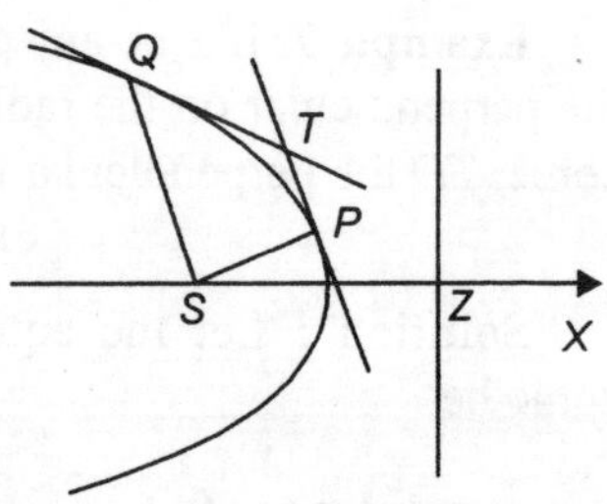

Fig.

Since, $P\,(SP, \alpha)$ lies on (i),

$$\therefore \quad \frac{l}{SP} = 1 + e \cos \alpha \qquad \text{...(ii)}$$

Since, $Q\,(SQ, \beta)$ lies on (i),

$$\therefore \quad \frac{l}{SQ} = 1 + e \cos \beta \qquad \text{...(iii)}$$

Multiplying (ii) and (iii), we get

$$\frac{l^2}{SP \cdot SQ} = (1 + e \cos \alpha)(1 + e \cos \beta) \qquad \text{...(iv)}$$

The equations of the tangents to (i) at P and Q are respectively,

$$\frac{l}{r} = e \cos \theta + \cos (\theta - \alpha) \qquad \text{...(v)}$$

and

$$\frac{l}{r} = e \cos \theta + \cos (\theta - \beta) \qquad \text{...(vi)}$$

Since, (v) and (vi) intersect at T.

$\therefore$ At T,

$$e\cos\theta + \cos(\theta-\alpha) = e\cos\theta + \cos(\theta-\beta)$$

$$\Rightarrow \qquad \cos(\theta-\alpha) = \cos(\theta-\beta)$$

Since, $\alpha \neq \beta$

$$\therefore \qquad \cos(\theta-\alpha) = \cos(\beta-\theta) \qquad [\because \cos(-\theta) = \cos\theta]$$

$$\Rightarrow \qquad \theta-\alpha = \beta-\theta$$

$$\Rightarrow \qquad 2\theta = \alpha+\beta$$

$$\Rightarrow \qquad \theta = \frac{\alpha+\beta}{2}$$

$\therefore$ From (v),

$$\frac{l}{ST} = e\cos\left(\frac{\alpha+\beta}{2}\right) + \cos\left(\frac{\alpha-\beta}{2}\right).$$

Squaring both sides, we get

$$\frac{l^2}{ST^2} = \left[e\cos\left(\frac{\alpha+\beta}{2}\right) + \cos\left(\frac{\alpha-\beta}{2}\right)\right]^2 \qquad \text{...(vii)}$$

$\therefore$ From (iv) and (vii),

$$l^2\left[\frac{1}{SP\cdot SQ} - \frac{1}{ST^2}\right]$$

$$= (1+e\cos\alpha)(1+e\cos\beta) - \left\{e\cos\left(\frac{\alpha+\beta}{2}\right) + \cos\left(\frac{\alpha-\beta}{2}\right)\right\}^2$$

$$= 1 + e(\cos\alpha+\cos\beta) + e^2\cos\alpha\cos\beta$$

$$-e^2\cos^2\frac{\alpha+\beta}{2} - 2e\cos\frac{\alpha+\beta}{2}\cos\frac{\alpha-\beta}{2} - \cos^2\frac{\alpha-\beta}{2}$$

$$= \left(1-\cos^2\frac{\alpha-\beta}{2}\right) + e^2$$

$$\left[\frac{1}{2}\{\cos(\alpha+\beta)+\cos(\alpha-\beta)\} - \cos^2\frac{\alpha+\beta}{2}\right]$$

$$= \sin^2\frac{\alpha-\beta}{2} + e^2$$

$$\left[\cos^2\frac{\alpha+\beta}{2} - \frac{1}{2} + \frac{1}{2} - \sin^2\frac{\alpha-\beta}{2} - \cos^2\frac{\alpha-\beta}{2}\right]$$

$$= (1-e^2)\sin^2\frac{\alpha-\beta}{2}$$

$$\Rightarrow \quad \frac{1}{SP \cdot SQ} - \frac{1}{ST^2} = \frac{1-e^2}{l^2}\sin^2\frac{\alpha-\beta}{2}$$

$$= \frac{b^2}{a^2 l^2}\sin^2\frac{\alpha-\beta}{2} \quad [\because b^2 = a^2(1-e^2)]$$

$$= \frac{b^2}{a^2}\cdot\frac{a^2}{b^4}\sin^2\frac{\alpha-\beta}{2}$$

$$= \frac{1}{b^2}\sin^2\frac{\alpha-\beta}{2} = \frac{1}{b^2}\sin^2\frac{PSQ}{2}.$$

Example 9. Show that the director circle of the conic,

$$\frac{l}{r} = 1 + e\cos\theta$$

is $\quad r^2(e^2-1) - 2elr\cos\theta + 2l^2 = 0.$

Solution : Director circle is defined as the locus of the point of intersection of two perpendicular tangents to a conic.

Let the equation of the conic be

$$\frac{l}{r} = 1 + e\cos\theta \qquad \text{...(i)}$$

Then the equations of the tangents at the points α and β of the conic are,

$$l/r = e\cos\theta + \cos\theta\,(\theta-\alpha) \qquad \text{...(ii)}$$

and $$l/r = e\cos\theta + \cos(\theta-\beta) \qquad \text{...(iii)}$$

Changing into cartesian form, we can easily see that

Slope of (ii) $= m_1 = -\dfrac{e+\cos\alpha}{\sin\alpha}$

and Slope of (iii) $= m_2 = -\dfrac{e+\cos\beta}{\sin\beta}$

If the tangents (ii) and (iii) are perpendiculars,

$$m_1 m_2 = -1$$

$$\Rightarrow \quad \left(-\frac{e+\cos\alpha}{\sin\alpha}\right)\left(-\frac{e+\cos\beta}{\sin\beta}\right) = -1$$

$$\Rightarrow \quad (e+\cos\alpha)(e+\cos\beta) + \sin\alpha\sin\beta = 0$$

$$\Rightarrow \quad e^2 + e(\cos\alpha+\cos\beta) + \cos(\alpha-\beta) = 0$$

$$\Rightarrow \quad e^2 + 2e\cos\frac{\alpha+\beta}{2}\cos\frac{\alpha-\beta}{2} + 2\cos^2\frac{\alpha-\beta}{2} - 1 = 0 \qquad \text{...(iv)}$$

At the point of intersection of (ii) and (iii)

$$\theta = \frac{\alpha+\beta}{2}$$

$\therefore$ From (ii),

$$\frac{l}{r} = e\cos\frac{\alpha+\beta}{2} + \cos\left(\frac{\alpha+\beta}{2} - \alpha\right)$$

[Substituting the value of θ]

$$\Rightarrow \quad \frac{l}{r} = e\cos\left(\frac{\alpha+\beta}{2}\right) + \cos\left(\frac{\alpha-\beta}{2}\right)$$

$$\Rightarrow \quad \frac{l}{r} - e\cos\frac{\alpha+\beta}{2} = \cos\frac{\alpha-\beta}{2}$$

$$\Rightarrow \quad \frac{l}{r} - e\cos\theta = \cos\frac{\alpha-\beta}{2} \qquad \text{...(v)}$$

Eliminating α and β from (iv) and (v), we have

$$e^2 + 2e\cos\theta\left(\frac{l}{r} - e\cos\theta\right) + 2\left(\frac{l}{r} - e\cos\theta\right)^2 - 1 = 0$$

$$\Rightarrow \quad r^2(e^2-1) - 2elr\cos\theta + 2l^2 = 0$$

which is the required equation of the director circle.

Example 10. At three points of the parabola $2a/r = 1 - \cos\theta$ whose vectorial angles are α, β and γ, tangents are drawn. Prove that the equation of the circle circumscribing the triangle formed by them is,

$$r = -a\operatorname{cosec}\frac{\alpha}{2}\operatorname{cosec}\frac{\beta}{2}\operatorname{cosec}\frac{\gamma}{2}\sin\left(\theta - \frac{\alpha+\beta+\gamma}{2}\right)$$

Hence, prove that it passes through the focus of the parabola.

Solution : The tangents at the points α, β, γ are respectively.

$$\frac{2a}{r} = -\cos\theta + \cos(\theta-\alpha) \qquad \text{...(i)}$$

$$\frac{2a}{r} = -\cos\theta + \cos(\theta-\beta) \qquad \text{...(ii)}$$

and

$$\frac{2a}{r} = -\cos\theta + \cos(\theta-\gamma) \qquad \text{...(iii)}$$

Solving (i) and (ii) for r and θ, we find that the tangents at the points α and β to the given parabola intersect at the point

$$\left(a\operatorname{cosec}\frac{\alpha}{2}\operatorname{cosec}\frac{\beta}{2}, \frac{\alpha+\beta}{2}\right).$$

Similarly, the point of intersection of (ii) and (iii) is,

$$\left(a \operatorname{cosec}\frac{\beta}{2}\operatorname{cosec}\frac{\gamma}{2}, \frac{\beta+\gamma}{2}\right)$$

and the point of intersection of (iii) and (i) is,

$$\left(a \operatorname{cosec}\frac{\gamma}{2}\operatorname{cosec}\frac{\alpha}{2}, \frac{\gamma+\alpha}{2}\right).$$

All these three points of intersection satisfy the equation,

$$r = -\,a \operatorname{cosec}\frac{\alpha}{2}\operatorname{cosec}\frac{\beta}{2}\operatorname{cosec}\frac{\gamma}{2}\sin\left\{\theta - \frac{\alpha+\beta+\gamma}{2}\right\}$$

which is clearly a circle.

If $\theta = \dfrac{\alpha+\beta+\gamma}{2}$, $r = 0$, i.e., the circle passes through the origin, i.e., the focus of the given parabola.

Example 11. Two equal ellipses of eccentricity e are placed with their axes at right angles and they have one focus S in common. If PQ be a common tangent, show that the angle PSQ is equal to $2\sin^{-1}\left(e/\sqrt{2}\right)$.

Solution: Let the equations of two equal ellipses having a common focus S and perpendicular axes be,

$$\frac{l}{r} = 1 + e\cos\theta \qquad \text{...(i)}$$

and

$$\frac{l}{r} = 1 + e\cos\left(\theta - \frac{\pi}{2}\right) \qquad \text{...(ii)}$$

Also, let the vectorial angles of P and Q be α and $\alpha + \beta$ respectively so that the common tangent PQ subtends and angle β at the common focus S.

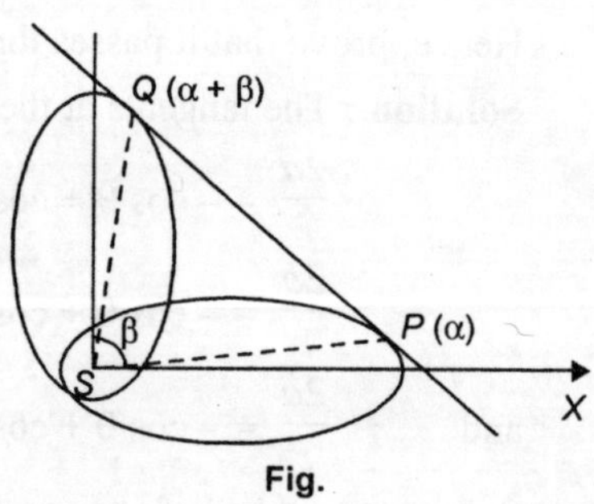

Fig.

Tangent at the point $P(\alpha)$ to the ellipse (i) is,

$$l/r = e\cos\theta + \cos(\theta - \alpha)$$
$$= (e + \cos\alpha)\cos\theta + \sin\alpha\sin\theta \qquad \text{...(iii)}$$

Tangent at the point $Q(\alpha + \beta)$ to the ellipse (ii) is,

$$\frac{l}{r} = e\cos\left(\theta - \frac{\pi}{2}\right) + \cos(\theta - \alpha + \beta)$$
$$= e\sin\theta + \cos(\theta - \alpha + \beta)$$
$$= \cos(\alpha + \beta)\cos\beta + [\sin(\alpha + \beta) + e]\sin\theta \quad \text{...(iv)}$$

Since, PQ is the common tangent, hence (iii) and (iv) must represent the same straight line.

Therefore, comparing the coefficients, we have

$$1 = \frac{e + \cos\alpha}{\cos(\alpha + \beta)} = \frac{\sin\alpha}{\sin(\alpha + \beta) + e}$$

These relations give

$$\cos\alpha - \cos(\alpha + \beta) = -e$$
$$\Rightarrow \quad 2\sin(\alpha + \beta/2)\sin\beta/2 = -e \quad \text{...(v)}$$
and
$$\sin(\alpha + \beta) - \sin\alpha = -e$$
$$\Rightarrow \quad 2\cos(\alpha + \beta/2)\sin\beta/2 = -e \quad \text{...(vi)}$$

Squaring and adding (v) and (vi), we get

$$4\sin^2\frac{\beta}{2} = 2e^2$$
$$\Rightarrow \quad 2\sin^2\beta/2 = e^2$$
$$\Rightarrow \quad \sin^2\beta/2 = e^2/2$$
$$\Rightarrow \quad \sin\beta/2 = e/\sqrt{2}$$
$$\Rightarrow \quad \beta/2 = \sin^{-1}\left(e/\sqrt{2}\right)$$
$$\Rightarrow \quad \beta = 2\sin^{-1}\left(e/\sqrt{2}\right).$$

EXERCISE 2 (B)

1. Show that the condition that the line $l/r = A\cos\theta + B\sin\theta$ may touch the conic $l/r = 1 + e\cos\theta$ is,
$$(A - e)^2 + B^2 = 1.$$
2. If the tangents at the points P and Q of a conic meet in a point T, prove that the vectorial angle of T is half of the sum of the vectorial angles of P and Q.
3. If the tangent at any point P on a conic whose focus is S meets the directrix in K, show that the angle PSK is a right angle.
4. Prove that the portion of the tangent intercepted between the conic and the directrix subtends a right angle at the corresponding focus.

5. PSP' is a focal chord of a conic. Prove that the tangents at P and P' intersect on the directrix.

6. Prove that the tangents drawn at the extremities of a focal chord of a parabola intersect at right angles.

7. If the tangents at the extremities of a chord through the focus S of a hyperbola meet the focal axis at P and P', prove that

$$\frac{1}{SP} + \frac{1}{SP'} = \text{constant}.$$

8. If the tangent from P to the conic $l/r = 1 + e \cos \theta$ subtends a fixed angle β at the focus S, prove that the locus of the mid-point of SP is conic of eccentricity $e \sec \beta$.

9. If the tangent at any point of an ellipse makes an angle α with its major axis and an angle β with the focal radius to the point of contact, show that

$$e \cos \alpha = \cos \beta.$$

10. A focal chord PSP' of an ellipse is inclined at an angle α to the major axis. Prove that the perpendicular from the focus on the tangent at P makes with the axis an angle

$$\tan^{-1}\left(\frac{\sin \alpha}{e + \cos \alpha}\right)$$

11. The tangents at three points P, Q, R of a parabola form a triangle ABC, show that

$$SP.SQ.SR = SA.SB.SC,$$

S being the focus of the parabola.

12. The tangents at P and Q of a conic meet in a point T, and S is the focus, prove that

$$ST^2 = SP.SQ$$

if the conic is a parabola.

13. The tangents at two points P and Q of a conic meet in T, and PQ subtends a constant angle 2α at the focus S, show that

$$\frac{1}{SP} + \frac{1}{PQ} - \frac{2\cos\alpha}{ST} = \text{constant}.$$

14. Prove that the equation of the auxiliary circle of the conic

$$l/r = 1 + e \cos \theta \text{ is},$$

$$r^2 (e^2 - 1) - 2elr \cos \theta + l^2 = 0.$$

Discuss the case when $e = 1$.

[***Hint.*** The auxiliary circle is the locus of the foot of the perpendicular from the focus upon any tangent to the conic.]

15. Show that a chord of a conic which subtends a constant angle at a focus of the conic envelopes (touches) another fixed conic.

16. Show that the two points having vectorial angles α and β on the conic $l/r = 1 + e \cos \theta$ will be ends of a diameter if

$$\tan \frac{\alpha}{2} \tan \frac{\beta}{2} = \frac{e+1}{e-1}.$$

17. If a chord of a conic section $l/r = 1 + e \cos \theta$ subtends a constant angle 2α at the focus, prove that the locus of the point where it meets the internal bisector of the angle 2α is the conic section

$$\frac{l \cos \alpha}{r} = 1 + e \cos \alpha \cos \theta.$$

18. A chord subtends a constant angle 2λ at the focus of the conic $l/r = 1 + e \cos \theta$, prove that the locus of the foot of the perpendicular on it from the focus if

$$r^2 (e^2 - \sec^2 \lambda) - 2elr \cos \theta + l^2 = 0.$$

19. If S is the focus and P and Q are two points on a conic, such that the angle PSQ is constant and equal to 2δ, prove that

(a) the locus of the point of intersection of the tangents at P and Q is a conic section whose focus is S.

(b) the line PQ always touches a conic whose focus is S.

20. P, Q, R are three points on the conic $l/r = 1 + e \cos \theta$, the focus S being the pole, SP and SR meet the tangents at Q in M and N so that $SM = SN = l$, prove that PR touches the curve

$$l/r = 1 + 2e \cos \theta$$

21. Find the tangent of conic $l/r = 1 + e \cos \theta$ at the etremities of its latus-rectum.

22. Show that locus of point of intersection of two tangents of parabola which cuts each other at constant angle α is the hyperbola

$$l/r = \cos \alpha + \cos \theta.$$

ANSWER

Exercise 2(B)

21. $\dfrac{l}{r} = e \cos \theta + \sin \theta.$

2.8 Polar

To find the equation of the polar of any point (r_1, θ_1) with respect to the conic $l/r = 1 + e\cos\theta$.

The polar of the point $T(r_1, \theta_1)$ is the chord of contact of tangents drawn from T to the conic.

Let the vectorial angles of the points of contact of the tangents from T be α and β respectively.

The equations of the tangents at the points $P(\alpha)$ and $Q(\beta)$ to the given conic are respectively

$$l/r = e\cos\theta + \cos(\theta - \alpha) \quad ...(1)$$

and

$$l/r = e\cos\theta + \cos(\theta - \beta) \quad ...(2)$$

Since, $T(r_1, \theta_1)$ is the point of intersection of these tangents, therefore, its coordinates must satisfy (1) and (2) both. So, we have

$$l/r_1 = e\cos\theta_1 + \cos(\theta_1 - \alpha) \quad ...(3)$$

and

$$l/r_1 = e\cos\theta_1 + \cos(\theta_1 - \beta) \quad ...(4)$$

Subtracting (4) from (3),

We have

$\cos(\theta_1 - \alpha) = \cos(\theta_1 - \beta)$.

Since, $\alpha \neq \beta$

$\therefore \cos(\theta_1 - \alpha) = \cos(\beta - \theta_1)$

$[\because \cos(-\theta) = \cos\theta]$

$\therefore \quad \theta_1 - \alpha = \beta - \theta_1$

$\Rightarrow \quad 2\theta_1 = \alpha + \beta$

$\Rightarrow \quad \theta_1 = \dfrac{\alpha + \beta}{2}$

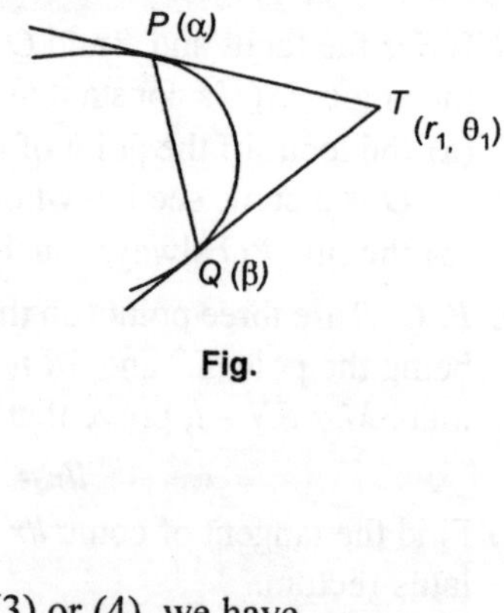

Fig.

Substituting this value of θ_1 in (3) or (4), we have

$$\frac{l}{r_1} = e\cos\theta_1 + \cos\left(\frac{\alpha+\beta}{2} - \alpha\right)$$

$$\Rightarrow \quad \frac{l}{r_1} - e\cos\theta_1 = \cos\left(\frac{\beta-\alpha}{2}\right) \quad ...(5)$$

Now, the equation of the joining $P(\alpha)$ and $Q(\beta)$ is,

$$\frac{l}{r} = e\cos\theta + \sec\left(\frac{\beta-\alpha}{2}\right)\cos\left(\theta - \frac{\alpha+\beta}{2}\right)$$

$$\Rightarrow \left(\frac{l}{r} - e\cos\theta\right)\cos\left(\frac{\beta-\alpha}{2}\right) = \cos\left(\theta - \frac{\alpha+\beta}{2}\right)$$

$$\Rightarrow \quad (l/r - e\cos\theta)\,(l/r_1 - e\cos\theta_1) = \cos(\theta - \theta_1)$$

which is the required equation of the polar of the point (r_1, θ_1).

Note. The point of intersection of the tangents i.e., $T(r_1, \theta_1)$ is called the pole of the PQ.

2.9 Equation of the Normal

To find the equation of the normal at the point 'α' of the conic

$$l/r = 1 + e\cos\theta.$$

Normal at any point of the conic is a line passing through that point and perpendicular to the tangent at that point.

The equation of the tangent at the point (α) of the given conic is,

$$l/r = e\cos\theta + \cos(\theta - \alpha)$$

$$\Rightarrow \quad l/r = (e + \cos\alpha)\cos\theta + \sin\alpha\sin\theta$$

$$\Rightarrow \quad l = (e + \cos\alpha)\, r\cos\theta + (\sin\alpha)\cdot r\sin\theta$$

$$\Rightarrow \quad l = (e + \cos\alpha)\, x + \sin\alpha \cdot y \qquad ...(1)$$

[changing into cartesian form]

Now, since $P(SP, \alpha)$ lies on

$$l/r = 1 + e\cos\theta$$

$$\therefore \quad \frac{l}{SP} = 1 + e\cos\alpha$$

$$\Rightarrow \quad SP = \frac{l}{1 + e\cos\alpha}$$

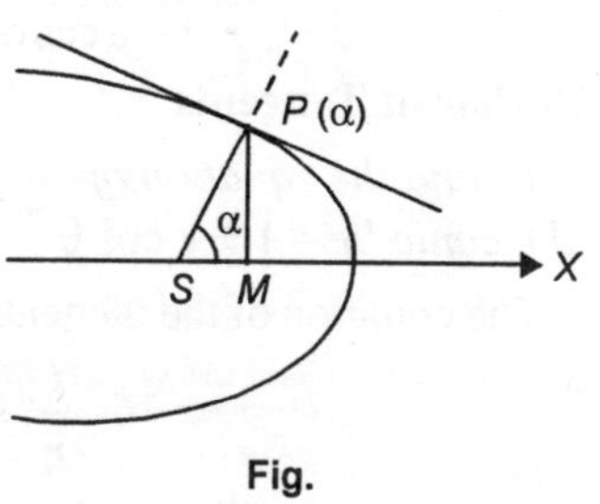

Fig.

Cartesian coordinates of P are,

$$(SP\cos\alpha,\ SP\sin\alpha)$$

i.e., $\left(\dfrac{l\cos\alpha}{1 + e\cos\alpha}, \dfrac{l\sin\alpha}{1 + e\cos\alpha}\right)$

Any line perpendicular to (1) is,

$$y(e + \cos\alpha) - x\sin\alpha = \lambda \qquad ...(2)$$

If it passes through P, then

$$\frac{l\sin\alpha}{1 + e\cos\alpha}(e + \cos\alpha) - \frac{l\cos\alpha\cdot\sin\alpha}{1 + e\cos\alpha} = \lambda$$

$$\Rightarrow \lambda = \frac{el \sin \alpha}{1 + e \cos \alpha}.$$

Substituting this value of λ in (2), the equation of normal in cartesian form is,

$$y(e + \cos \alpha) - x \sin \alpha = \frac{el \sin \alpha}{1 + e \cos \alpha}.$$

Transforming this into polar form, we have

$$r \sin \theta \, (e + \cos \alpha) - r \cos \theta \sin \alpha = \frac{el \sin \alpha}{1 + e \cos \alpha}.$$

$$\Rightarrow \qquad \frac{e \sin \alpha}{1 + e \cos \alpha} \cdot \frac{l}{r} = e \sin \theta + \sin (\theta - \alpha)$$

which is the required polar equation of the normal.

Note. (i) If the equation of the conic is,

$$l/r = 1 + e \cos (\theta - \gamma),$$

then the equation of the normal at the point α is,

$$\frac{e \sin (\alpha - \gamma)}{1 + e \cos (\alpha - \gamma)} \frac{l}{r} = e \sin (\theta - \gamma) + \sin (\theta - \alpha).$$

(ii) If conic is $l/r = 1 - e \cos \theta$, then equation of normal is,

$$-\frac{l}{r} \frac{e \sin \alpha}{1 - e \cos \alpha} = \sin (\theta - \alpha) - e \sin \theta.$$

2.10 Pair of Tangents

To find the equation of the pair of tangents from a point (r_1, θ_1) to the conic $l/r = 1 + e \cos \theta$.

The equation of the tangent at any point 'α' of the given conic is,

$$\frac{l}{r} = e \cos \theta + \cos (\theta - \alpha) \qquad ...(1)$$

The tangent will pass through (r_1, θ_1) if,

$$\frac{l}{r_1} = e \cos \theta_1 + \cos (\theta_1 - \alpha) \qquad ...(2)$$

The required equation of the pair of tangents from the point (r_1, θ_1) to the given conic will be obtained by eliminating α between (1) and (2).

So, from (1) and (2), we have

$$\left[\left(\frac{l}{r} - e \cos \theta\right)^2 - 1\right]\left[\left(\frac{l}{r_1} - e \cos \theta_1\right)^2 - 1\right]$$

$$= [\cos^2(\theta - \alpha) - 1]\,[\cos^2(\theta_1 - \alpha) - 1]$$
$$= [-\sin^2(\theta - \alpha)]\,[-\sin^2(\theta_1 - \alpha)]$$
$$= \sin^2(\theta - \alpha)\sin^2(\theta_1 - \alpha) \qquad ...(3)$$

Also, $\left(\dfrac{l}{r} - e\cos\theta\right)\left(\dfrac{l}{r_1} - e\cos\theta_1\right) - \cos(\theta - \theta_1)$

$$= \cos(\theta - \alpha)\cos(\theta_1 - \alpha) - \cos(\theta - \theta_1)$$
$$= \frac{1}{2}\cdot 2\cos(\theta - \alpha)\cos(\theta_1 - \alpha) - \cos(\theta - \theta_1)$$
$$= \frac{1}{2}\cdot[\cos(\theta + \theta_1 - 2\alpha) + \cos(\theta - \theta_1)] - \cos(\theta - \theta_1)$$
$$= \frac{1}{2}\cdot[\cos(\theta + \theta_1 - 2\alpha) - \cos(\theta - \theta_1)]$$
$$= \frac{1}{2}\cdot[-2\sin(\theta - \alpha)\sin(\theta_1 - \alpha)]$$
$$= -\sin(\theta - \alpha)\sin(\theta_1 - \alpha). \qquad ...(4)$$

Hence, from (3) and (4), the required equation of the pair of tangents from the point (r_1, θ_1) is,

$$\left[\left(\frac{l}{r} - e\cos\theta\right)^2 - 1\right]\left[\left(\frac{l}{r_1} - e\cos\theta_1\right)^2 - 1\right]$$
$$= \left[\left(\frac{l}{r} - e\cos\theta\right)\left(\frac{l}{r_1} - e\cos\theta_1\right) - \cos(\theta - \theta_1)\right]^2$$

$$\Rightarrow \qquad (S^2 - 1)(S_1^{\,2} - 1) = [SS_1 - \cos(\theta - \theta_1)]^2$$

where $S = \dfrac{l}{r} - e\cos\theta$ and $S_1 = \dfrac{l}{r_1} - e\cos\theta_1$.

ILLUSTRATIVE EXAMPLES

Example 1. Given the focus and directrix of a conic, show that the polar of a given point with respect to it passes through a fixed point.

Solution : Let the equation of the conic be,

$$\frac{l}{r} = 1 + e\cos\theta. \qquad ...(i)$$

Let S be the given focus and ZM the directrix.

Let $SZ = k$ (given).

Then by definition of a conic,

$$l/k = e \Rightarrow l = ek \qquad \text{...(ii)}$$

$\therefore$ (i) becomes,

$$\frac{ek}{r} = 1 + e\cos\theta. \qquad \text{...(iii)}$$

Let (r_1, θ_1) be the given point.

$\therefore$ The polar of (r_1, θ_1) with respect to (iii) is,

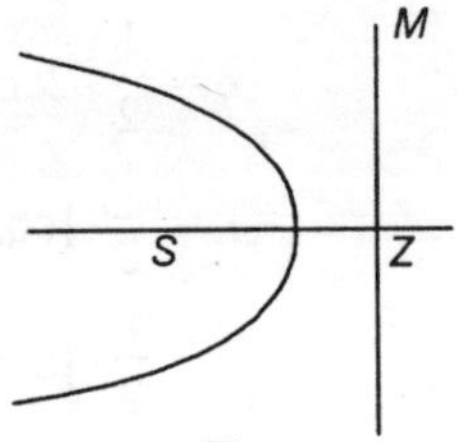

Fig.

$$\left(\frac{ek}{r} - e\cos\theta\right)\left(\frac{ek}{r_1} - e\cos\theta_1\right)$$

$$= \cos(\theta - \theta_1)$$

$$\Rightarrow e^2\left(\frac{k}{r} - \cos\theta\right)\left(\frac{k}{r_1} - \cos\theta_1\right) = \cos(\theta - \theta_1) \qquad \text{...(iv)}$$

This line, for all values of e, passes through the point of intersection of the lines,

$$k/r - \cos\theta = 0 \qquad \text{...(v)}$$

and $$\cos(\theta - \theta_1) = 0 \qquad \text{...(vi)}$$

Equation (vi) gives

$$\theta = \theta_1 + (\pi/2).$$

$\therefore$ From (v),

$$r = \frac{k}{\cos\theta}$$

$$= \frac{k}{\cos(\theta_1 + \pi/2)} = -\frac{k}{\sin\theta_1}$$

$$= -k\,\text{cosec}\,\theta_1.$$

Hence, the polar of (r_1, θ_1) passes through the fixed point,

$$(-k\,\text{cosec}\,\theta_1, \theta_1 + \pi/2).$$

Example 2. Find the locus of the pole of a chord of the conic $l/r = 1 + e\cos\theta$ which subtends a constant angle 2α at the focus.

Solution : Let the chord of the conic,

$$l/r = 1 + e\cos\theta \qquad \text{...(i)}$$

be PQ.

Let the vectorial angles of P and Q be $\lambda - \alpha$ and $\lambda + \alpha$ respectively, so that

$$\angle QSP = 2\alpha.$$

The equation of the tangents to (i) at the points P and Q are respectively,

$$l/r = e \cos \theta + \cos (\theta - \lambda + \alpha) \quad \text{...(ii)}$$

and $$l/r = e \cos \theta + \cos (\theta - \lambda - \alpha) \quad \text{...(iii)}$$

Since, the pole of the chord PQ is the point of intersection of the tangents at P and Q, hence the locus of the pole is obtained by eliminating λ between (ii) and (iii).

Subtracting (iii) from (ii), we get

$$\cos (\theta - \lambda + \alpha) = \cos (\theta - \lambda - \alpha).$$

But since, $\alpha \neq 0$.

Therefore, $\cos (\theta - \lambda + \alpha) = \cos (-\theta + \lambda + \alpha)$

$[\because \cos(-\theta) = \cos\theta]$

$$\Rightarrow \quad \theta - \lambda + \alpha = -\theta + \lambda + \alpha$$

$$\Rightarrow \quad 2\lambda = 2\theta$$

$$\Rightarrow \quad \lambda = \theta.$$

Substituting it in (ii), we get

$$l/r = e \cos \theta + \cos \alpha$$

$$\Rightarrow \quad \frac{l \cdot \sec \alpha}{r} = 1 + (e \sec \alpha) \cos \theta.$$

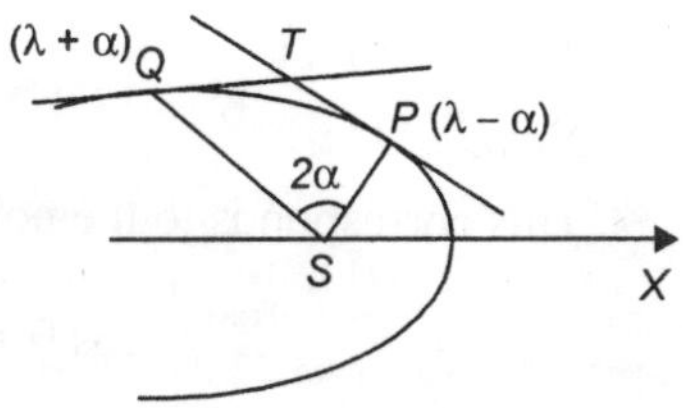

Fig.

This is the required locus which is a conic of latus rectum $2l \sec \alpha$ and eccentricity $e \sec \alpha$, focus being S.

Example 3. If the normal at L, one of the extremities of the latus rectum of the conic $l/r = 1 + e \cos \theta$ meets the curve again at Q; show that

$$SQ = \frac{1 + 3e^2 + e^4}{1 + e^2 - e^4} \cdot l.$$

Solution : The given conic is,

$$l/r = 1 + e \cos \theta \quad \text{...(i)}$$

Let LSL' be the latus rectum of the conic (i). Then,

$$L \to (l, \pi/2).$$

Equation of normal at L is,

$$\frac{el}{r} = e \sin\theta + \sin\left(\theta - \frac{\pi}{2}\right)$$

$$\Rightarrow \frac{el}{r} = e\sin\theta - \cos\theta \quad ...(ii)$$

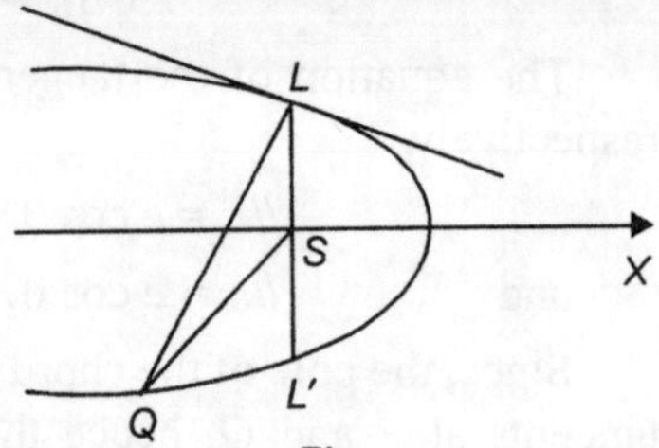

Fig.

It meets the curve again at Q.

Therefore, solving (i) and (ii), we get

$$e(1 + e\cos\theta) = e\sin\theta - \cos\theta$$

$$\Rightarrow \quad e + (e^2 + 1)\cos\theta = e\sin\theta.$$

Squaring both sides,

$$e^2 + (e^2+1)^2\cos^2\theta + 2e(e^2+1)\cos\theta = e^2\sin^2\theta$$
$$= e^2(1 - \cos^2\theta)$$
$$= e^2 - e^2\cos^2\theta$$

$$\Rightarrow \quad \cos^2\theta\,[e^2 + (e^2+1)^2] + 2e(e^2+1)\cos\theta = 0$$

$$\Rightarrow \cos\theta\,[\cos\theta\,\{e^2 + (e^2+1)^2\} + 2e(e^2+1)] = 0$$

$$\Rightarrow \quad \cos\theta\,[\cos\theta\,\{e^4 + 3e^2 + 1\} + 2e(e^2+1)] = 0$$

$$\Rightarrow \quad \cos\theta = 0 \text{ or } \cos\theta = -\frac{2e(e^2+1)}{e^4+3e^2+1}$$

Now, $\cos\theta = 0 \Rightarrow \theta = \pi/2.$

This corresponds to the point L. Hence, for the other point Q,

$$\cos\theta = \frac{-2e(e^2+1)}{e^4+3e^2+1}$$

Now, Q lies on (i)

$$\therefore \quad \frac{l}{SQ} = 1 + e\left\{\frac{-2e(e^2+1)}{e^4+3e^2+1}\right\}$$

$$\Rightarrow \quad \frac{l}{SQ} = \frac{e^4+3e^2+1-2e^4-2e^2}{e^4+3e^2+1}$$

$$\Rightarrow \quad \frac{l}{SQ} = \frac{1+e^2-e^4}{e^4+3e^2+1}$$

$$\Rightarrow \quad SQ = \frac{l(1+3e^2+e^4)}{1+e^2-e^4}.$$

Example 4. If the normals at α, β, γ, δ on the conic $l/r = 1 + e\cos\theta$ meet at (R, ϕ), prove that

(1) $\tan\dfrac{\alpha}{2}\tan\dfrac{\beta}{2}\tan\dfrac{\gamma}{2}\tan\dfrac{\delta}{2} + \left(\dfrac{1+e}{1-e}\right)^2 = 0$

(2) $\alpha + \beta + \gamma + \delta - 2\phi$ = an odd multiple of π.

Solution : Let the given conic be,

$$l/r = 1 + e\cos\theta \quad \text{...(i)}$$

The equation of the normal at any point λ is,

$$\frac{e\sin\lambda}{1-e\cos\lambda}\cdot\frac{l}{r} = e\sin\theta + \sin(\theta-\lambda)$$

$$= (e + \cos\lambda)\sin\theta - \sin\lambda\cos\theta$$

$$\Rightarrow \quad el\sin\lambda = (e+\cos\lambda)(1+e\cos\lambda)\,r\sin\theta - \sin\lambda\,(1+e\cos\lambda)\,r\cos\theta \quad \text{...(ii)}$$

Now, we know that

$$\sin\lambda = \frac{2\tan\lambda/2}{1+\tan^2\lambda/2} = \frac{2t}{1+t^2} \text{ where } t = \tan\frac{\lambda}{2}$$

and

$$\cos\lambda = \frac{1-\tan^2\lambda/2}{1+\tan^2\lambda/2} = \frac{1-t^2}{1+t^2}$$

Substituting these values, (ii) becomes,

$$r\sin\theta\left[e+\frac{1-t^2}{1+t^2}\right]\left[1+e\frac{1-t^2}{1+t^2}\right] - r\cos\theta\,\frac{2t}{1+t^2}\left[1+e\frac{1-t^2}{1+t^2}\right] - el\cdot\left[\frac{2t}{1+t^2}\right] = 0$$

$$\Rightarrow r\sin\theta\left[\frac{e(1+t^2)+(1-t^2)}{1+t^2}\right]\left[\frac{1+t^2+e(1-t^2)}{1+t^2}\right] - 2r\cos\theta\left[\frac{(1+t^2)+e(1-t^2)}{1+t^2}\right]\frac{t}{1+t^2} - el\cdot\frac{2t}{1+t^2} = 0$$

$$\Rightarrow r\sin\theta\left[\frac{(1+e)^2-(1-e)^2t^4}{(1+t^2)^2}\right] - 2r\cos\theta\left[\frac{(1+e)\,t+(1-e)\,t^3}{(1+t^2)^2}\right] - \frac{2elt}{1+t^2} = 0$$

$$\Rightarrow r \sin\theta\,[(1+e)^2-(1-e)^2t^4]-2r\cos\theta\,[(1+e)\,t+(1-e)\,t^3] \\ -2elt\,(1+t^2)=0$$

$$\Rightarrow t^4\,[(1-e)^2 r\sin\theta]+2t^3\,[el+(1-e)\,r\cos\theta]+2t\,[el+(1+e) \\ r\cos\theta]-(1+e)^2\,r\sin\theta=0.$$

It passes through the point (R, ϕ), therefore,

$t^4\,[(1-e)^2 R\sin\phi]+2t^3\,[el+(1-e)\,R\cos\phi]+2t\,[el+(1+e)\,R$ $\cos\phi]-(1+e)^2 R\sin\phi\ =0.$

This is a biquadratic equation in t. Hence, it will give four values of t say t_1, t_2, t_3, t_4. Let these four values be $\tan\alpha/2$, $\tan\beta/2$, $\tan\gamma/2$ and $\tan\delta/2$ respectively.

$$\therefore \qquad S_1=\Sigma t_1=\frac{-2\left[el+(1-e)\,R\cos\phi\right]}{(1-e)^2 R\sin\phi}$$

$$S_2=\Sigma t_1t_2=0$$

$$S_3=\Sigma t_1t_2t_3=\frac{-2\left[el+(1+e)\,R\cos\phi\right]}{(1-e)^2 R\sin\phi}$$

$$S_4=\Sigma t_1t_2t_3t_4=-\left(\frac{1+e}{1-e}\right)^2$$

$$\Rightarrow \qquad \tan\frac{\alpha}{2}\tan\frac{\beta}{2}\tan\frac{\gamma}{2}\tan\frac{\delta}{2}+\left(\frac{1+e}{1-e}\right)^2=0$$

which proves result (i)

$$\text{Now, } \tan\left(\frac{\alpha}{2}+\frac{\beta}{2}+\frac{\gamma}{2}+\frac{\delta}{2}\right)=\frac{S_1-S_3}{1-S_2+S_4}$$

$$=-\cot\phi$$

$$=\tan\left(\frac{\pi}{2}+\phi\right)$$

$$\Rightarrow \qquad \frac{\alpha}{2}+\frac{\beta}{2}+\frac{\gamma}{2}+\frac{\delta}{2}=n\pi+\frac{\pi}{2}+\phi$$

$$\Rightarrow \qquad \alpha+\beta+\gamma+\delta=(2n+1)\,\pi+2\phi$$

$$\Rightarrow \qquad \alpha+\beta+\gamma+\delta-2\phi=(2n+1)\,\pi$$

$$= \text{an odd multiple of } \pi.$$

Example 5. Show that the equation to the circle which passes through the focus and touches the curve $l/r = 1 + e\cos\theta$ at the point $\theta = \alpha$, is,

$$r\,(1+e\cos\alpha)^2=l\cos(\theta-\alpha)+el\cos(\theta-2\alpha).$$

Solution : Let S is focus of conic

$$\frac{l}{r} = 1 + e\cos\theta \qquad \text{...(i)}$$

Let the centre of the circle which touches conic (i) be (α, β). Let P be point on conic whose vectorial angle is α.

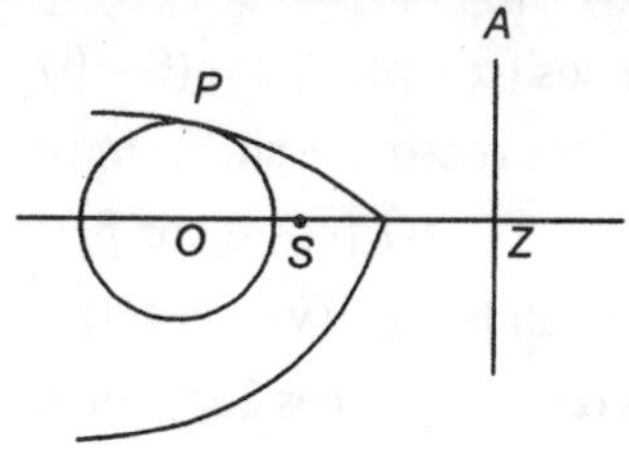

Fig.

Equation of circle through S having centre at O, is

$$r = 2a\cos(\theta - \beta) \qquad \text{...(ii)}$$

P, a point whose vectorial angle is α, is common to (i) and (ii). Hence,

$$\frac{l}{1+e\cos\alpha} = 2a\cos(\alpha - \beta) \qquad \text{...(iii)}$$

At the point of contact P, the two curves (i) and (ii) have common tangent, so normal to conic (i) will pass through centre O of circle (ii).

Normal to conic (i) at point P is,

$$\frac{l}{r}\frac{e\sin\alpha}{1+e\cos\alpha} = \sin(\theta - \alpha) + e\sin\theta \qquad \text{...(iv)}$$

Point (a, β) lies on it.

$$\therefore \quad \frac{l}{a}\frac{e\sin\alpha}{1+e\cos\alpha} = \sin(\beta - \alpha) + e\sin\beta \qquad \text{...(v)}$$

Eliminating a between (iii) and (v)

$$\frac{el\sin\alpha}{1+e\cos\alpha} = \frac{l\{e\sin\beta + \sin(\beta-\alpha)\}}{2(1+e\cos\alpha)\cos(\theta-\alpha)}$$

$$\sin(\beta - \alpha) + e\sin\beta = 2e\sin\alpha\cos(\alpha - \beta)$$

$$\Rightarrow \quad \sin(\beta - \alpha) + e\sin\beta = e\{\sin(2\alpha - \beta) + \sin\beta\}$$

$$\Rightarrow \quad \sin(\beta - \alpha) = e\sin(2\alpha - \beta)$$

$$\Rightarrow -\sin\alpha\cos\beta + \cos\alpha\sin\beta = e(\sin 2\alpha\cos\beta - \cos 2\alpha\sin\beta)$$

$$\Rightarrow \cos\beta(\sin\alpha + e\sin 2\alpha) = \sin\beta(\cos\alpha + e\cos 2\alpha)$$

$$\Rightarrow \qquad \frac{\sin\beta}{\sin\alpha + e\sin 2\alpha} = \frac{\cos\beta}{\cos\alpha + e\cos 2\alpha} \qquad ...(vi)$$

Eliminating a from (ii) and (iii)

$$\frac{\cos(\theta-\beta)}{\cos(\alpha-\beta)} = \frac{r(1+e\cos\alpha)}{l}$$

$\Rightarrow r(1 + e\cos\alpha)\cos(\alpha - \beta) = l\cos(\theta - \beta)$

$\Rightarrow r(1 + e\cos\alpha)\{\cos\alpha\cos\beta + \sin\alpha\sin\beta\}$

$= l\{\cos\theta\cos\beta + \sin\theta\sin\beta\}$ ...(vii)

Putting the value of tan β from (v) and (vi)

$r(1 + e\cos\alpha)\{\cos\alpha(\cos\alpha + e\cos 2\alpha) + \sin\alpha(\sin\alpha + e\sin 2\alpha)\}$

$= l\{\cos\theta(\cos\alpha + e\cos 2\alpha) + \sin\theta(\sin\alpha + e\sin 2\alpha)\}$

$\Rightarrow r(1 + e\cos\alpha)\{\cos^2\alpha + e\cos\alpha\cos 2\alpha + \sin^2\alpha +$

$e\sin\alpha\sin 2\alpha\} = l\{\cos\theta\cos\alpha + e\cos\theta\cos 2\alpha$

$+ \sin\theta\sin\alpha + e\sin\theta\sin 2\alpha\}$

$\Rightarrow r(1 + e\cos\alpha)^2 = l\cos(\theta - \alpha) + l\,e\cos(\theta - 2\alpha).$

EXERCISE 2 (C)

1. If the chord PQ of a conic subtends a right angle at the focus of the conic, prove that the locus of the pole of PQ is a conic having the same focus.
2. If the polars of (r_1, α) and (r_2, β) with respect to a conic $1 + \cos\theta$ are at right angles, prove that

$$\frac{l^2}{r_1 r_2} + \sin\alpha\sin\beta = 0.$$

3. If the normals at three points α, β, γ of the parabola

$$l/r = 1 + \cos\theta$$

meet in the point (e, ϕ), prove that $2\phi = \alpha + \beta + \gamma$.

4. If the normals at three points α, β, γ of the parabola,

$$2a/r = 1 - \cos\theta \text{ or } r = a\operatorname{cosec}^2\theta/2$$

meet in a point whose vectorial angle is δ, prove that

$$2\delta = \alpha + \beta + \gamma - \pi.$$

5. If the normals at three points P, Q, R on a parabola $2a/r = 1 + \cos\theta$ meet in a point, O and S be the focus, prove that

$$SP \,.\, SQ \,.\, SR = a \,.\, SO^2.$$

6. Two points P and Q are taken on a parabola whose vectorial angles are supplementary. Show that the locus of the point of intersection of the normals at P and Q is a parabola.

OBJECTIVE TYPE EXAMPLE

Example. In any conic semi latus rectum is,

(i) arithmetic mean (ii) harmonic mean

(iii) geometric mean (iv) none of above.

between the segment of any focal chord.

Solution : Let conic is,

$$\frac{l}{r} = 1 + e\cos\theta \qquad \text{...(i)}$$

Let PSQ is any focal chord, where S is pole.

If vectorial angle of P is α then vectorial angle of Q is $\pi + \alpha$, points P and Q are on conic.

$$\frac{l}{SP} = 1 + e\cos\alpha \qquad \text{...(ii)}$$

and
$$\frac{l}{SQ} = 1 + e\cos(\pi+\alpha) = 1 - e\cos\alpha \qquad \text{...(iii)}$$

Now,
$$\frac{l}{SP} + \frac{l}{SQ} = 2$$

$$\Rightarrow \quad \frac{l}{SP} + \frac{l}{SQ} = \frac{2}{l}$$

$$\Rightarrow \quad \frac{\frac{1}{SP} + \frac{1}{SQ}}{2} = \frac{1}{l}$$

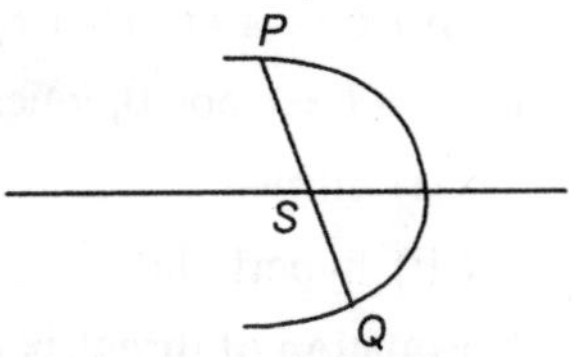

Fig.

It shows Harmonic mean of SP and SQ is l. Hence, choice (ii) is correct.

OBJECTIVE EXERCISE

1. Directrix of conic $l/r = 1 + e\cos\theta$ is,

(i) $l/r = 1 + e\cos\theta$ (ii) $l/r = e\cos\theta$

(iii) $l/r = 1 - e\cos\theta$ (iv) none of these.

2. Equation $\frac{1}{r} = 2\sin^2\frac{\theta}{2}$ represents,

(i) ellipse (ii) hyperbola

(iii) parabola (iv) straight line.

3. Equation of normal of conic $l/r = 1 + e\cos\theta$ at point $\theta = \alpha$ is,

(i) $\dfrac{el\sin\alpha}{r(1+e\cos\alpha)} = -e\sin\theta + \sin(\theta-\alpha)$

(ii) $\dfrac{el\sin\alpha}{r(1+e\cos\alpha)} = e\sin\theta + \sin(\theta-\alpha)$

(iii) $\dfrac{el\sin\alpha}{r(1+e\cos\alpha)} = \sin\theta - e\sin(\theta-\alpha)$

(iv) $\dfrac{el\sin\alpha}{r(1+e\cos\alpha)} = -e\sin\theta - \sin(\theta-\alpha)$

4. Equation of tangent of conic $l/r = 1 + e\cos\theta$ at a point $\theta = \alpha$ is,

(i) $l/r = e\cos(\theta-\alpha)$ (ii) $l/r = e\cos(\theta-\alpha) + \cos\theta$

(iii) $l/r = e\cos\theta + \cos(\theta-\alpha)$ (iv) $l/r = 1 + e\cos(\theta-\alpha)$

5. Equation of tangent of conic $l/r = 1 + e\cos(\theta - r)$ at point α is,

(i) $l/r = e\cos(\theta-\alpha) + \cos(\theta-r)$

(ii) $l/r = e\cos(\theta+r) + \cos(\theta-\alpha)$

(iii) $l/r = e\cos(\theta-r) + \cos(\theta+\alpha)$

(iv) $l/r = e\cos(\theta+r) + \cos(\theta+\alpha)$.

6. $l/r = 1 + e\cos\theta$, where $e > 1$ is equation of,

(i) ellipse (ii) parabola

(iii) hyperbola (iv) none of these.

7. Equation of directrix of conic $l/r = 1 + e\cos\theta$ corresponding to focus other than pole is,

(i) $\dfrac{l}{r} = \dfrac{1-e^2}{1+e^2}e\cos\theta$ (ii) $\dfrac{l}{r} = \dfrac{1+e^2}{1-e^2}e\sin\theta$

(iii) $\dfrac{l}{r} = \dfrac{1-e^2}{1+e^2}e\cos\theta$ (iv) $\dfrac{e}{r} = -\dfrac{1-e^2}{1+e^2}e\cos\theta$

8. Eccentricity of rectangular hyperbola is,

(i) 1 (ii) 2

(iii) $\sqrt{2}$ (iv) < 1

9. Equation $\dfrac{l}{r} = 1 + e\cos\theta$ is parabola if,

(i) $e = 0$ (ii) $e > 1$

(iii) $e < 1$ (iv) $e = 1$

10. On the conic $l/r = 1 + e \cos \theta$ vectorial angle of P and Q are $\alpha - \beta$ and $\alpha + \beta$, equation of chord PQ is

(i) $l/r = e \cos \theta + \cos \{\theta - (\alpha + \beta)\}$

(ii) $l/r = e \cos \theta + \sec \beta \cos (\theta - \alpha)$

(iii) $l/r = e \sin \theta + \cos \{\theta - (\alpha + \beta)\}$

(iv) $l/r = e \sin \theta + \sec \beta \cos (\theta + \alpha)$.

ANSWERS

Objective Exercise

1. (ii)	**2.** (iii)	**3.** (ii)	**4.** (iii)	**5.** (iii)
6. (iii)	**7.** (iv)	**8.** (iii)	**9.** (iv)	**10.** (ii)

3

General Equation of Second Degree

3.1 Introduction

It has been observed that all the curves were obtained, originally, by cutting a cone (not necessarily a right circular cone) by a plane in various ways. Hence these curves are known as conic sections.

3.2 Definition

A conic section or conic is the locus of a point P which moves under the condition that its distance from a fixed point S is in a constant ratio to its perpendicular distance from a fixed straight line, all being in the same plane.

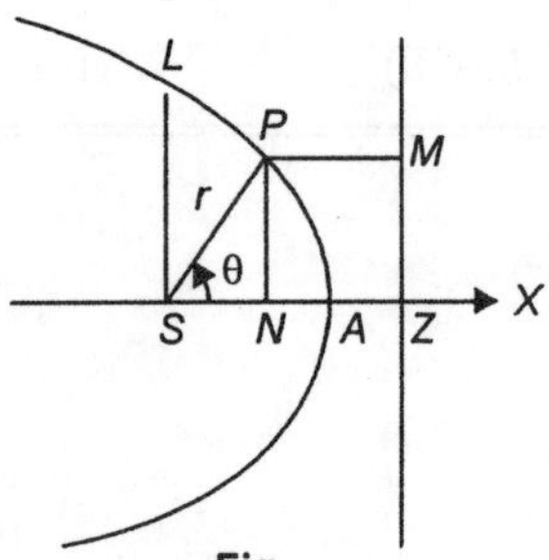

Fig.

Important terms

(i) **Focus :** The afore-said fixed point S is called the focus of the conic section. Mainly there are two real foci of a conic to which we are concerned.

(ii) **Directrix :** The fixed straight line, mentioned in 3.2, is called the directrix of the conic section.

For every focus there will be a corresponding directrix.

(iii) **Eccentricity :** The constant ratio, mentioned in 3.2, is called the eccentricity of the conic and it is denoted by e.

(iv) **Axis :** The straight line, passing through the focus and perpendicular to the directrix, is called the **axis** of the conic.

(v) **Vertices :** The point of intersection, of the conic and the axis, is called the **vertex** of the conic.

(vi) **Centre :** The point, which bisects every chord of the conic, passing through it is called the centre of the conic. This point bisects the line joining the two foci of the conic.

3.3 General Equation of the Conic

Let, $S(\alpha, \beta)$ be the focus, $Ax + By + C = 0$ be the equation of the directrix. Let the coordinates of any point P on the conic be (x, y). Then by the definition,

$$\frac{SP}{PM} = e$$

$$\Rightarrow \quad SP^2 = e^2 PM^2, \qquad \ldots(1)$$

where, M is the foot of the perpendicular from P upon the directrix. The cartesian equation of the conic section will be, by (1).

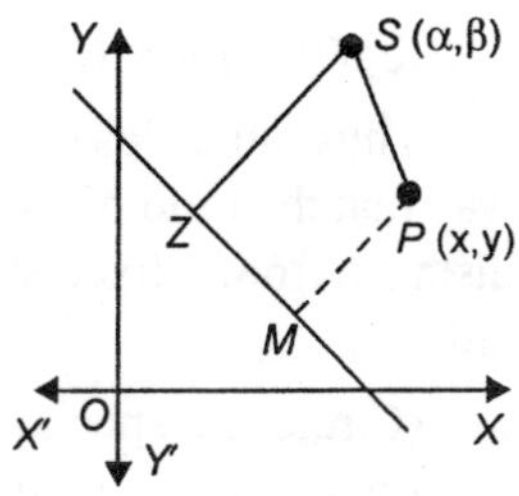

Fig.

$$(x-\alpha)^2 + (y-\beta)^2 = e^2 \frac{(Ax+By+C)^2}{A^2+B^2},$$

e, being the eccentricity. Simplifying the above equation we get it in the form,

$$ax^2 + 2hxy + by^2 + 2gx + 2fy + c = 0$$

or briefly, $f(x, y) = 0$.

3.4 Particular forms of the Conic

The nature of the conic section depends upon the position of the focus S and the value of the eccentricity e;

Mainly we face two cases :

Case 1 : *When the focus S does not lie on the directrix*

The most simple case of equation of the conic can be taken when the directrix is taken as the axis of y and the focus is taken on the axis of x. Let $(\alpha, 0)$ be the coordinates of focus S, then, the equation of the conic in general form can be written as

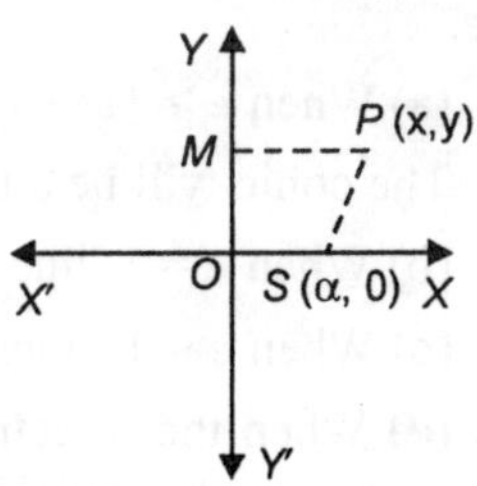

Fig.

$$(x-\alpha)^2 + y^2 = e^2x^2 \qquad \ldots(1)$$

Hence, the nature of the conic in this case will be depending upon the value of e, as in the following cases :

(a) ***When e = 0.***

Then the equation (1) will represent a circle as a circle is the locus of a point which moves in such a way that its distance from a fixed point is always constant.

The fixed point is called the centre of the circle and the fixed distance is called its radius. The two foci coincide with its centre while the two directrices lie at infinity.

(b) ***When e = 1. The conic is a parabola.***

Thus, a parabola is the locus of a point which moves in such a way that its distance from a fixed point, called the focus, is equal to its distance from a fixed line, called its directrix. Thus, one set of focus, directrix and its centre will lie at infinity.

(c) ***When e < 1. The conic is an ellipse.***

Thus, an ellipse is the locus of a point which moves in such a way that the ratio of its distance from a fixed point called focus to its distance from a fixed straight line called directrix is always less than unity.

(Circle is a special case of the ellipse).

(d) ***When e > 1. The conic is called a hyperbola.***

It can be defined as in the case of ellipse. Here the distance of the moving point from the focus will bear a constant ratio with its distance from the directrix and this ratio will be greater than unity.

Case 2 : ***When the focus lies on the directrix, i.e.*** α ***of equation (1) of case 1 is zero***

In this case the equation reduces in the form

$$y^2 = (e^2 - 1)\, x^2$$

$$\Rightarrow \qquad y = \pm\sqrt{e^2 - 1}\cdot x$$

which will represent a pair of straight lines depending upon the value of e.

(a) When $e > 1$

The conic will be a pair of real and distinct straight lines.

(b) When $e = 1$, lines will be coincident.

(c) When $e < 1$, lines will be imaginary.

(d) When the directrix with focus on it is at infinity, the conic will be a pair of parallel lines.

3.5 General Equation of Second Degree

In 3.3, we got the equation of the conic section in the form,

$$ax^2 + 2hxy + by^2 + 2gx + 2fy + c = 0 \qquad \text{...(1)}$$

in which there are six arbitrary constants named as a, b, c, f, g, h

which can be reduced to any other five arbitrary constants by dividing it by anyone non-zero constant of the above six arbitrary constants.

If $a \neq 0$, then the equation (1) of this section is a second degree equation in general form.

Hence,

$$\boldsymbol{f(x, y) \equiv ax^2 + 2hxy + by^2 + 2gx + 2fy + c = 0}$$

is called the general equation of second degree.

Note. In the general equation of second degree, i.e.,

$$ax^2 + 2hxy + by^2 + 2gx + 2fy + c = 0$$

$$a = \text{coefficient of } x^2$$

$$b = \text{coefficient of } y^2$$

$$c = \text{free constant term}$$

$$f = \frac{1}{2} \times \text{the coefficient of } y$$

$$g = \frac{1}{2} \times \text{the coefficient of } x$$

$$h = \frac{1}{2} \times \text{the coefficient of } xy.$$

$$\Delta \equiv abc + 2fgh - af^2 - bf^2 - ch^2$$

is called the **Discriminant** of the general requation of second degree. It is easy to verify that

$$\Delta \equiv \begin{vmatrix} a & h & g \\ h & b & f \\ g & f & c \end{vmatrix}$$

3.6 To Show that the General Equation of Second Degree Always Represents a Conic

As we know that general equation of second degree, referred to rectangular axes, is given by,

$$f(x, y) \equiv ax^2 + 2hxy + by^2 + 2gx + 2fy + c = 0 \quad ...(1)$$

Let the coordinate axes be turned through an angle θ without shifting the origin, to get the equation (1) free from the term xy. Then, in standard form, x and y of the above equation will be transformed in the forms, $(x \cos \theta - y \sin \theta)$ and $(x \sin \theta + y \cos \theta)$ respectively. The new transformed equation will thus be,

$$a (x \cos \theta - y \sin \theta)^2 + 2h (x \cos \theta - y \sin \theta) (x \sin \theta + y \cos \theta)$$
$$+ b (x \sin \theta + y \cos \theta)^2 + 2g (x \cos \theta - y \sin \theta)$$
$$+ 2f (x \sin \theta + y \cos \theta) + c = 0 \quad ...(2)$$

The coefficient of xy in (2) will be

$2(b-a)\sin\theta\cos\theta + 2h(\cos^2\theta - \sin^2\theta)$

i.e. $(b-a)\sin 2\theta + 2h\cos 2\theta$.

In order to eliminate the term xy from (1), choose θ so that the coefficient of xy in (2) becomes zero.

i.e. $\quad (b-a)\sin 2\theta + 2h\cos 2\theta = 0$

$$\Rightarrow \quad \tan 2\theta = \frac{2h}{a-b}$$

which gives a real value of 2θ for all real values of a, b and h. For this value of θ, equation (2) will be transformed in the form,

$$Ax^2 + By^2 + 2Gx + 2Fy + C = 0 \qquad \ldots(3)$$

where, A, B, G, F and C are known constants now.

Case 1 : *If neither A nor B is Zero*

The equation (3) can be reduced to the form

$$A\left(x+\frac{G}{A}\right)^2 + B\left(Y+\frac{F}{B}\right)^2 = \frac{G^2}{A} + \frac{F^2}{B} - C$$

$$= K \text{ (say)}.$$

Transferring the origin to $\left(-\frac{G}{A}, -\frac{F}{B}\right)$ equation (3) becomes,

$$Ax^2 + By^2 = K \qquad \ldots(4)$$

(a) **If $K = 0$, $\Delta = 0$**

Equation (4) represents a pair of straight lines, real or imaginary according as A and B are of opposite signs or of the same sign.

(b) **If $K \neq 0$**

Equation (4) can be written as,

$$\frac{x^2}{(K/A)} + \frac{y^2}{(K/B)} = 1$$

which is the equation to an **ellipse** or a **hyperbola** according as K/A and K/B are of the same sign or of opposite sign.

If K/A and K/B both are negative, then no real values of x and y will satisfy the equation, but on account of its form, the curve will be regarded as an **imaginary ellipse.**

(c) **If $K \neq 0$ and $A = B$**

The equation (4) will represent a **circle**.

(d) **When $K \neq 0$ and $A = -B$**

The equation (4) will represent a **rectangular hyperbola.**

Case 2 : *If A or B is Zero*

Suppose that $A = 0$. (If $B = 0$ in place of A, the procedure will be the similar one giving a similar result.)

In this case the equation (3) will be reduced to,

$$B\left(y+\frac{F}{B}\right)^2 = -2Gx + \frac{F^2}{B} - C \qquad \text{...(5)}$$

(a) **If $G = 0$**

Equation (5) will represent two parallel straight lines which will be coincident if $\left(\frac{F^2}{B} - C\right)$ vanishes.

(b) **If $G \neq 0$**

The equation (5) can be further reduced to the form

$$\left(y+\frac{F}{B}\right)^2 = -\frac{2G}{B}\left(x - \frac{F^2}{2BG} + \frac{C}{2G}\right),$$

Transferring the origin to $\left(\frac{F^2}{2BG} - \frac{C}{2G}, -\frac{F}{B}\right)$

the above equation becomes

$$y^2 = -\left(\frac{2G}{B}\right)x, \text{ which represents a parabola.}$$

Thus, we find that the general equation of second degree always represents a conic section.

Note. If, $f(x, y) \equiv ax^2 + 2hxy + by^2 + 2gx + 2fy + c = 0$ be the general equation of second degree, then it will represent different conic sections under different conditions given as below :

No.	Conditions	Nature of the Conic
1.	$\Delta = 0,\ h^2 \neq ab$	pair of straight lines
2.	$\Delta = 0,\ h^2 = ab$	pair of parallel straight lines
3.	$\Delta = 0,\ a + b = 0$	pair of perpendicular straight lines
4.	$\Delta \neq 0,\ a = b,\ h = 0$	Circle
5.	$\Delta \neq 0,\ h^2 = ab$	Parabola
6.	$\Delta \neq 0,\ h^2 < ab$	Ellipse
7.	$\Delta \neq 0,\ h^2 > ab$	Hyperbola
8.	$\Delta \neq 0,\ h^2 > ab,\ a + b = 0$	Rectangular hyperbola

In this section, we are mainly concerned with circle, parabola, ellipse and hyperbola.

3.7 Centre

To find the coordinates of the centre of the conic represented by general equation of second degree :

Let the equation of second degree be,

$$ax^2 + 2hxy + by^2 + 2gx + 2fy + c = 0 \qquad ...(1)$$

Let (x', y') be the coordinates of the centre of the conic represented by (1). Shifting the origin at this centre and keeping the axes parallel to original axes, the equation (1) will be transformed to

$$a(x + x')^2 + 2h(x + x')(y + y') + b(y + y')^2$$
$$+ 2g(x + x') + 2f(y + y') + c = 0 \qquad ...(2)$$

Since (x', y') is the centre which has become the new origin for (2) hence, any chord passing through it will be bisected at this point (Definition of centre). This shows that if, referred to new axes, (x_1, y_1) lies on (2), then $(-x_1, -y_1)$ will also lie on it as x', y' is the origin (new). Since, $(-x_1, -y_1)$ and (x_1, y_1) both will satisfy (2), hence it must not contain any first degree term in it. Hence coefficients of x and y in (2) must vanish.

Therefore,

$$ax' + hy' + g = 0 \qquad ...(3)$$
$$hx' + by' + f = 0 \qquad ...(4)$$

which on solving for x', y', gives

$$x' = \frac{hf - bg}{ab - h^2}, y' = \frac{gh - af}{ab - h^2}$$

Hence, coordinates, of the centre of the conic (1), referred to old axes, will be

$$\left(\frac{hf - bg}{ab - h^2}, \frac{gh - af}{ab - h^2}\right)$$

Note. If $ab - h^2 = 0$, $gh - af$ or $hf - bg = 0$, then

$$\frac{a}{h} = \frac{h}{b} = \frac{g}{f} \text{ and}$$

the coordinates of the centre become meaningless i.e. the conic has no centre.

The above conditions with (3) and (4) give a single straight line.

In this case any point satisfying (3) or (4) may be regarded as the centre of the conic.

3.8 Equation of the Conic Section Referred to the Centre

Let the equation of the conic be

$$f(x, y) \equiv ax^2 + 2hxy + by^2 + 2gx + 2fy + c = 0 \qquad \text{...(1)}$$

of which (x', y') be the centre. Then we have,

$$ax' + hy' + g = 0 \qquad \text{...(2)}$$

and
$$hx' + by' + f = 0 \qquad \text{...(3)}$$

When the origin be transferred to (x', y') and axes being parallel to old one, the new equation will become,

$$ax^2 + 2hxy + by^2 + c' = 0$$

where $c' = ax'^2 + 2hx'y' + by'^2 + 2gx' + 2fy' + c$ ...(4)

Multiplying the equations (2) and (3) by x' and y' respectively and adding, we get

$$ax'^2 + 2hx'y' + by'^2 + gx' + fy' = 0$$

which reduces (4) in the form

$$c' = gx' + fy' + c \qquad \text{...(5)}$$

Putting the values of x', y' obtained in 3.7, we get

$$c' = g\left(\frac{hf - bg}{ab - h^2}\right) + f\left(\frac{gh - af}{ab - h^2}\right) + c$$

$$= \frac{abc + 2fgh - af^2 - bg^2 - ch^2}{ab - h^2}$$

$$= \frac{\Delta}{ab - h^2}$$

Thus, the equation of the conic section referred to centre as origin, becomes

$$ax^2 + 2hxy + by^2 + \frac{\Delta}{ab - h^2} = 0$$

Note. If, $ax^2 + 2hxy + by^2 + 2gx + 2fy + c = 0$

be the equation of the conic section, then,

$$\Delta = \begin{vmatrix} a & h & g \\ h & b & f \\ g & f & c \end{vmatrix}$$

If (x', y') be the coordinates of the centre of the conic then,

$$ax' + hy' + g = 0$$
$$hx' + by' + f = 0$$

give the centre and

$$gx' + fy' + c = \frac{\Delta}{ab^2 - h^2}.$$

3.9 Equation of Asymptotes

We know that the equation of the asymptote of a conic, whose equation is reduced to its standard form, differs by a constant term only.

This property remains unaltered when the equation of the conic is changed back referred to its original axes.

Hence, if

$$ax^2 + 2hxy + by^2 + 2gx + 2fy + c = 0 \qquad ...(1)$$

be the equation of the conic, then the equation of its **asymptotes** will be,

$$ax^2 + 2hxy + by^2 + 2gx + 2fy + c + \lambda = 0 \qquad ...(2)$$

Since asymptotes are straight lines, hence, (2) will represent a pair of straight lines. Therefore, applying the condition that a second degree equation representing a pair of straight lines, we get

$$\begin{vmatrix} a & h & g \\ h & b & f \\ g & f & c+\lambda \end{vmatrix} = 0$$

which gives

$$\lambda\,(ab - h^2) + \Delta = 0 \qquad ...(3)$$

where $$\Delta = \begin{vmatrix} a & h & g \\ h & b & f \\ g & f & c \end{vmatrix}$$

Putting the value of λ from (3) in (2), we get the equation of the asymptotes of the conic (1) as,

$$ax^2 + 2hxy + by^2 + 2gx + 2fy + c - \frac{\Delta}{ab^2 - h^2} = 0$$

Note 1. The conic represented by the equation,

$$(a_1x + b_1y + c_1)\,(a_2x + b_2y + c_2) + K = 0$$

is a hyperbola whose asymptotes are the lines given by

$$a_1x + b_1y + c_1 = 0$$

and

$$a_2x + b_2y + c_2 = 0$$

Note 2. The equation of the conjugate hyperbola differs from the equation of the asymptotes by a constant which is equal to the difference between the equations of the conic and the asymptote.

Hence, the equation of the conjugate hyperbola of (1) will be

$$ax^2 + 2hxy + by^2 + 2gx + 2fy + c - \frac{2\Delta}{ab - h^2} = 0.$$

3.10 Nature of the Conic

Let,

$$ax^2 + 2hxy + by^2 + 2gx + 2fy + c = 0 \qquad ...(1)$$

be the equation of the conic, then the equation of its asymptotes will be

$$ax^2 + 2hxy + by^2 + 2gx + 2fy + c - \frac{\Delta}{ab - h^2} = 0.$$

Therefore,

$$ax^2 + 2hxy + by^2 = 0 \qquad ...(2)$$

will represent straight lines parallel to these asymptotes.

Hence,

Case 1 : If $h^2 - ab > 0$, these straight lines are real and distinct. Hence the conic (1) will be a hyperbola.

Case 2 : If $h^2 - ab = 0$, then the second degree terms of (1) will form a perfect square, hence the conic represented by (1) will be a parabola.

Case 3 : If $h^2 - ab < 0$, the lines represented by (2) will be imaginary. Hence (1) will represented as ellipse.

Thus in a nut-shell, the general equation of second degree will represent a hyperbola, a parabola or an ellipse according as

$h^2 - ab >$, = , or < 0 unless $\Delta = 0$. In this case it will represent two straight lines.

Particular Case : If $h^2 - ab > 0$ and $a + b = 0$, the asymptotes of the hyperbola are at right angle, hence, the conic will be a rectangular hyperbola.

3.11 To Find the Lengths and the Positions of the Axes of the Conic

Let the equation of the conic referred to its centre as origin and axes parallel to old axes, be

$$ax^2 + 2hxy + by^2 = 1 \quad ...(1)$$

and let the equation of a circle, whose centre be at the centre of (1) and its radius be r, referred to the centre of (1) as origin, be

$$x^2 + y^2 = r^2$$

$$\Rightarrow \quad (x^2 + y^2)/r^2 = 1 \quad ...(2)$$

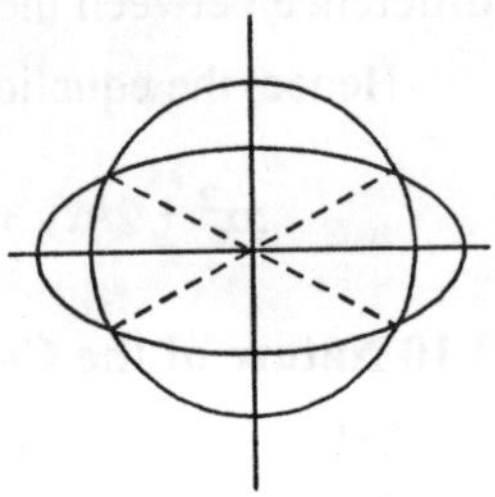

Fig.

Then the equation of the straight lines joining the points of intersection of (1) and (2) with the origin i.e., the common centre, will be obtained by making (1) homogeneous of second degree with the help of (2). Such straight lines will be given by

$$ax^2 + 2hxy + by^2 = \frac{x^2 + y^2}{r^2}$$

$$\therefore \left(a - \frac{1}{r^2}\right) x^2 + 2hxy + \left(b - \frac{1}{r^2}\right) y^2 = 0 \quad ...(3)$$

These straight lines represented by (3) will be the common diameters of (1) and (2) shown by the dotted lines.

These diameters will coincide, if r be such that,

$$\left(a - \frac{1}{r^2}\right)\left(b - \frac{1}{r^2}\right) = h^2 \quad ...(4)$$

which is a quadratic equation in r^2 giving the square of lengths of semi-axes of the conic (1). Let these be r_1^2 and r_2^2. Then, r_1 and r_2 will be the lengths of semi-axes of the conic.

Multiplying (3) by $\left(a - \frac{1}{r^2}\right)$ we get,

$$\left(a - \frac{1}{r^2}\right)^2 x^2 + 2h\left(a - \frac{1}{r^2}\right) xy + \left(a - \frac{1}{r^2}\right)\left(b - \frac{1}{r^2}\right) y^2 = 0$$

$$\Rightarrow \quad \left(a - \frac{1}{r^2}\right) x^2 + 2h\left(a - \frac{1}{r^2}\right) xy + h^2 y^2 = 0 \quad \text{[by (4)]}$$

$$\therefore \qquad \left\{\left(a - \frac{1}{r^2}\right)x + hy\right\}^2 = 0$$

$$\Rightarrow \qquad \left(a - \frac{1}{r^2}\right) x + hy = 0 \qquad ...(5)$$

which will be the equation of the either axes of the conic (1). In turn if we put r_1^2 or r_2^2 given by (4), we get the equation of either axes referred to new system of axes with centre as origin.

Using the transformation formula i.e. $x - x'$ and $y - y'$ for x and y of (5), we get the equation of the axes in reference to original axes, (x', y') being the coordinates of the centre.

Note. If the transformed equation of the conic referred to centre as origin, be

$$ax^2 + 2hxy + by^2 = K,$$

then the squares of the semi-axes are given by,

$$\left(a - \frac{K}{r^2}\right)\left(b - \frac{K}{r^2}\right) = h^2$$

Consequently, the equations of the axes referred to new origin will be,

$$\left(a - \frac{K}{r_1^2}\right)x + hy = 0$$

and $$\left(a - \frac{K}{r_2^2}\right)x + hy = 0$$

3.12 Eccentricity, Foci and Directrices

If out of the two roots equation (4) of the previous article r_1^2 be the larger one, say $r_1^2 = R^2$ and $r_2^2 = r^2$ then, R^2 and r^2 both being positive give the conic as an ellipse. Consequently its eccentricity

$$e = \sqrt{\left(1 - \frac{r^2}{R^2}\right)}$$

The foci are points on the major axis at a distance eR i.e. $\sqrt{R^2 - r^2}$ from the centre (x', y'). Hence, their coordinates referred to original axes will be

$$\left\{x' \pm \left(\sqrt{R^2 - r^2}\right)\cos\theta, y' \pm \left(\sqrt{R^2 - r^2}\right)\sin\theta\right\}$$

where $$\tan\theta = -\frac{\left(a-\frac{1}{R^2}\right)}{h}$$ i.e.

the gradient of the major axis.The directrices are the lines perpendicular to the major axis at a distance $R/e = R^2/\sqrt{R^2-r^2}$ from the centre. Hence, their equations referred to old axes will be

$$(x-x')\cos\theta + (y-y')\sin\theta = \pm R^2/\sqrt{R^2-r^2}.$$

Note. The same formulae hold also in case when the conic is a hyperbola only with the difference that r^2 is negative.

3.13 Tracing of Conics

In case, the conic be an ellipse or a hyperbola, then find out its

(i) Centre,

(ii) Equation of the conic referred to centre as origin,

(iii) Lengths and positions of the axes,

(iv) Eccentricity, foci, directrices,

(v) Points where the conic meets the coordinate axes,

(vi) Equation of asymptotes in case of hyperbola, and

(vii) Rough sketch of the conic.

This will be called the tracing of the conic.

ILLUSTRATIVE EXAMPLES

Example 1. Trace the conic,

$$17x^2 - 12xy + 8xy^2 + 46x - 28y + 17 = 0.$$

Also find the equations of its directrices.

Solution: Let (x', y') be the coordinates of its centre, then we have

$$17x' - 6y' + 23 = 0$$

and $$-6x' + 8y' - 14 = 0$$ (See note of 3.8)

Solving above equations, we get

$$x' = -1 \text{ and } y' = 1$$

Therefore, coordinates of the centre are $(-1, 1)$.

Also, $c' = gx' + fy' + c$

$= -23 - 14 + 17$

$= -20$

Hence, equation of the conic, referred to origin at centre and axes being parallel to original axes, will be

$$17x^2 - 12xy + 8y^2 + c' = 0$$

$$\Rightarrow 17x^2 - 12xy + 8y^2 - 20 = 0$$

$$\Rightarrow \frac{1}{20}(17x^2 - 12xy + 8y^2) = 1$$

Hence, the length of its axes will be given by,

$$\left(\frac{17}{20} - \frac{1}{r^2}\right)\left(\frac{8}{20} - \frac{1}{r^2}\right) = \left(-\frac{6}{20}\right)^2$$

$$\Rightarrow \quad (17r^2 - 20)(2r^2 - 5) = 9$$

$$\Rightarrow \quad 25r^4 - 125r^2 + 100 = 0$$

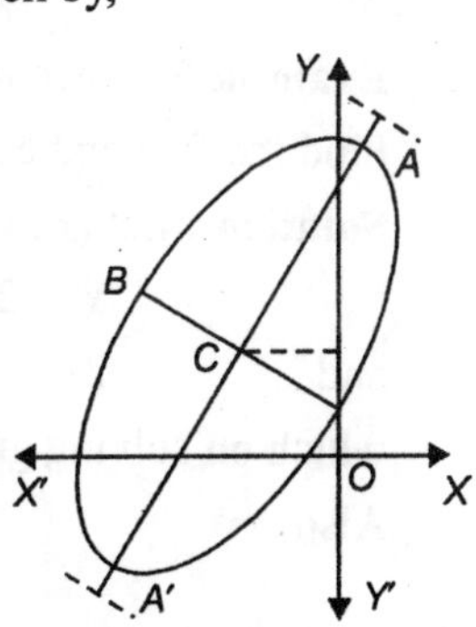

Fig.

which is quadratic in r^2. Hence solving it we get, $r^2 = 4$ or 1. Since, both the values of r^2 are positive, the conic is an ellipse, its semi-axes are of lengths $R = 2$ and $r = 1$, and the gradient of the major axis is equal to

$$-\left(a - \frac{1}{R^2}\right)\Big/h = -\left(\frac{17}{20} - \frac{1}{4}\right)\Big/\left(\frac{-3}{10}\right)$$

$$= 2.$$

Where the ellipse meets the axis of x, we have

$$17x^2 + 46x + 17 = 0$$

which gives,

$$x = -\frac{23}{17} \pm \frac{4}{17}\sqrt{15}$$

i.e. $\quad = -0.44$ and -2.26 nearly.

Also, where it meets the axis of y, we get

$$8y^2 - 28y + 17 = 0$$

which on solving gives,

$$y = 2.72 \text{ and } 0.78\text{, nearly.}$$

Thus, the shape of the ellipse will be as shown in the diagram.

The directrices are perpendicular to the major axes and at a distance $\frac{R^2}{\sqrt{R^2 - r^2}}$ i.e. $\frac{4}{\sqrt{3}}$ from the centre.

Since the gradient of the major axis is 2, the equation of the directrices will be, $(x - x') \cos \theta + (y - y') \sin \theta = \pm R^2/\sqrt{R^2 - r^2}$

(By 3.12)

$$(x+1)\left(\frac{1}{\sqrt{5}}\right) + (y-1)\left(\frac{2}{\sqrt{5}}\right) = \pm\frac{4}{\sqrt{3}}$$

i.e. $$x + 2y - 1 \pm 4\sqrt{5/3} = 0.$$

Example 2. Trace the conic, $x^2 + 4xy + y^2 - 2x + 2y - 6 = 0$. Find the foci and eccentricity of the conic.

Solution : Let (x', y') be the centre of the conic. Then,

$$x' + 2y' - 1 = 0$$

and $$2x' + y' + 1 = 0$$

which on solving give the centre at $(-1, 1)$.

Also, $$c' = gx' + fy' + c$$
$$= 1 + 1 - 6$$
$$= -4$$

Hence, equation of the conic, referred to centre as origin and axes parallel to old axes, will be

$$ax^2 + 2hxy + by^2 + c' = 0$$

$$\Rightarrow \quad x^2 + 4xy + y^2 - 4 = 0$$

$$\Rightarrow \quad \frac{1}{4}(x^2 + 4xy + y^2) = 1$$

and the lengths of axes are given by

$$\left(\frac{1}{4} - \frac{1}{r^2}\right)\left(\frac{1}{4} - \frac{1}{r^2}\right) = \left(\frac{1}{2}\right)^2$$

$$\Rightarrow \quad (r^2 - 4)^2 = 4r^4$$

$$\Rightarrow \quad r^2 - 4 = \pm 2r^2$$

$$\Rightarrow \quad r^2 = \frac{4}{3} \text{ or } -4$$

Since the two values of r^2 are of opposite signs, the conic is a hyperbola.

$$e = \sqrt{1 - \frac{(-4)}{4/3}} = 2$$

and the gradient of the transverse axis

$$= \tan\theta$$

$$= \left(\frac{1}{4} - \frac{3}{4}\right) \Big/ \left(-\frac{1}{2}\right)$$

$$= 1 \qquad \therefore \theta = 45°$$

Foci will lie on it at a distance of eR from the centre i.e. at a distance $2\sqrt{4/3}$ from $(-1, 1)$.

Hence their coordinates will be

$$\left(-1 \pm 2\sqrt{4/3}\cos\theta, 1 \pm 2\sqrt{4/3}\sin\theta\right)$$

i.e. $$\left(-1 \pm 2\frac{\sqrt{4}}{3}\cdot\frac{1}{\sqrt{2}}, 1 \pm 2\frac{\sqrt{4}}{3}\cdot\frac{1}{\sqrt{2}}\right)$$

i.e. $$\left(-1 + 2\frac{\sqrt{2}}{3}, 1 + 2\frac{\sqrt{2}}{3}\right)$$

and $$\left(-1 - 2\frac{\sqrt{2}}{3}, 1 - 2\frac{\sqrt{2}}{3}\right)$$

The points, where the hyperbola cuts the axis of x, are given by,

$$x^2 - 2x - 6 = 0$$

i.e. $$x = \frac{2 \pm \sqrt{4 + 24}}{2}$$

$$= \frac{2 \pm 2\sqrt{7}}{2}$$

$$= 1 \pm 2.6$$

$$= 3.6 \text{ and } -1.6$$

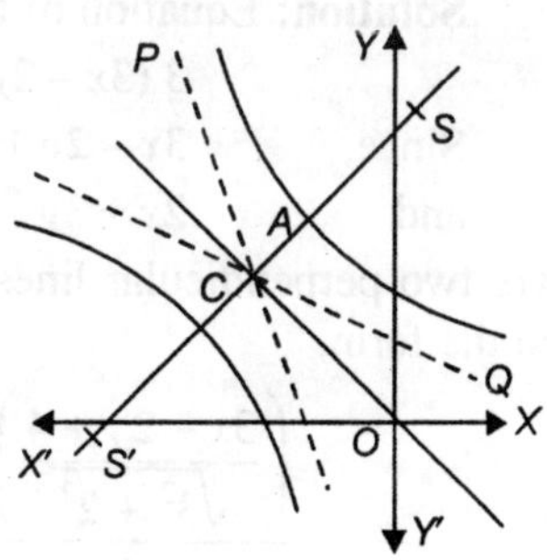

Fig.

Similarly, where it meets the axis of y, we get

$$y^2 + 2y - 6 = 0$$

$$\therefore \qquad y = \frac{-2 \pm \sqrt{4 + 24}}{2}$$

$$= -1 \pm \sqrt{7}$$
$$= -1 \pm 2.6$$
$$= 1.6 \text{ and } -3.6.$$

Hence the shape of the hyperbola will be as given below :

Asymptotes : Here in the question,

$$\Delta = abc + 2fgh - af^2 - bg^2 - ch^2$$
$$= (1)(1)(-6) + 2(1)(-1)(2) - 1(1)^2 - 1 - (-1)^2 - (-6)(2)^2$$
$$= -6 - 4 - 1 - 1 + 24$$
$$= 12$$

and $\quad ab - h^2 = (1)(1) - (2)^2$
$$= -3$$

Hence, equation of the asymptotes will be given by,

$$ax^2 + 2hxy + by^2 + 2gx + 2fy + c - \frac{\Delta}{ab - h^2} = 0$$

$$\Rightarrow \quad x^2 + 4xy + y^2 - 2x + 2y - 6 - \frac{12}{-3} = 0$$

$$\Rightarrow \quad x^2 + 4xy + y^2 - 2x + 2y - 2 = 0$$

which gives

$$x + (2 - \sqrt{3})\,y + (\sqrt{3} - 1) = 0$$

and $\quad x + (2 + \sqrt{3})\,y - (\sqrt{3} + 1) = 0$

Example 3. Trace the conic

$$3(3x - 2y + 4)^2 + 2(2x + 3y - 5)^2 = 39.$$

Solution: Equation of the given conic is

$$3(3x - 2y + 4)^2 + 2(2x + 3y - 5)^2 = 39 \qquad \ldots\text{(i)}$$

Since, $\quad 3x - 2y + 4 = 0$

and $\quad 2x + 3y - 5 = 0,$

are two perpendicular lines, we can arrange the given equation (i) in the form,

$$\frac{\left(\dfrac{3x - 2y + 4}{\sqrt{3^2 + 2^2}}\right)^2}{1} + \frac{\left(\dfrac{2x + 3y - 5}{\sqrt{2^2 + 3^2}}\right)^2}{3/2} = 1$$

Putting, $X = \dfrac{2x + 3y - 5}{\sqrt{13}}$, $Y = \dfrac{3x - 2y + 4}{\sqrt{13}}$

The above equation reduces in the form,

$$\frac{X^2}{3/2}+\frac{Y^2}{1}=1$$

which represents an ellipse whose major and minor axes are given by the equations,

$$Y=0$$

and $$X=0$$

i.e. $$3x-2y+4=0 \quad \text{...(ii)}$$

and $$2x+3y-5=0 \quad \text{...(iii)}$$

Solving these equations, we get the coordinates of the centre as, $\left(-\frac{2}{13},\frac{23}{13}\right)$.

Length of semi-major and semi-minor axes are $\sqrt{3/2}$ and 1.

i.e. 1.58 and 1 approx.

Tracing : Take the centre $C\left(-\frac{2}{13},\frac{23}{13}\right)$

Through C draw lines ACA' included at an angle $\tan^{-1}(3/2)$ to the X-axis and mark of $CA = CA'$ = 1.58 along this line.

Draw another line BCB' at right angles to ACA' and cut $CB = CB' = 1$ along it.

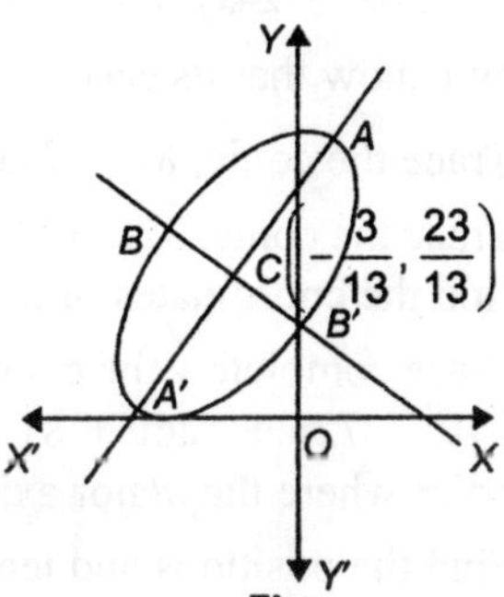

Fig.

Also, Put $y = 0$ in (i) to get the point of intersection of the conic and the axis of x, we get

$$35x^2+32x+59=0$$

which gives imaginary values of x showing that the conic does not intersect the axis of x.

Similarly putting $x = 0$ in (i) we get the values of y, where the conic meets Y axis. For that, we get

$$30y^2-108y+59=0,$$

giving $y = 2{\cdot}9$ and $\cdot 7$ (nearly). Thus, the shape of the curve will be as shown in the figure.

EXERCISE 3 (A)

1. Find the centre and the lengths of the axes of the conic,

$$14x^2 - 4xy + 11y^2 - 44x - 58y + 71 = 0.$$

2. Find the equation of the asymptotes of the conic,

$$3x^2 - 2xy - 5y^2 + 7x - 9y = 0$$

and also find the equation of the conic which has the same asymptotes and which passes through the point (2, 2).

3. Find the positions and magnitudes of the axes of the conic whose equation is, $x^2 - 5xy + y^2 + 8x - 20y + 14 = 0$.

Trace the Conics :

4. $21x^2 - 6xy + 29y^2 + 6x + 58y - 151 = 0$.

5. $x^2 - xy + 2y^2 - x - 4y - 2 = 0$.

6. $x^2 - 3xy - y^2 + 10x - 10y + 21 = 0$.

7. $3x^2 + 8xy - 3y^2 - 40x + 20y + 50 = 0$.

8. $2(3x - 4y - 1)^2 + 3(4x + 3y - 3)^2 = 8$.

9. $(x + y - 1)(x - 7y - 9) = 2$

10. Trace the conic,

$$34x^2 + 24xy + 41y^2 + 48x + 14y - 108 = 0$$

and show that its eccentricity is $1/\sqrt{2}$.

11. Trace the conic, $8x^2 - 24xy + 15y^2 + 48x - 48y = 0$.

12. Trace the conic, $32x^2 + 52xy - 7y^2 - 64x - 52y - 148 = 0$ and also find the coordinates of its foci.

13. Trace completely the conic given by the equation, $36x^2 + 24xy + 29y^2 - 72y + 126y + 81 = 0$. Also find the coordinates of the point where the major axis cuts the conic.

14. Find the positions and lengths of the axes of the conic,

$$x^2 - 4xy - 2y^2 + 10x + 4y = 0,$$

Explain the steps taken. Trace the conic.

15. Show that the product of the semi-axes of the ellipse whose equation is, $x^2 - xy + 2y^2 - 2x - 6y + 7 = 0$ is $2/\sqrt{7}$ and that the equation of the axes is,

$$x^2 - y^2 - 2xy + 8y - 8 = 0.$$

16. Show that the equation of the hyperbola which has $3x - 4y + 7 = 0$ and $4x + 3y + 1 = 0$ for its asymptotes and passes through the origin is $12x^2 - 7xy - 12y^2 + 31x + 7y = 0$. Trace the curve.

ANSWERS

Exercise 3(A)

1. (2, 3); $2\sqrt{6}$, 4.
2. $3x^2 - 2xy - 5y^2 + 7x - 9y + 2 = 0$,
   $3x^2 - 2xy - 5y^2 + 7x - 9y + 20 = 0$.
3. Transverse axis : $x + y + 4 = 0$, length $2\dfrac{\sqrt{2}}{7}$; conjugate axis : $x - y + 4 = 0$, length $2\sqrt{2/3}$.
4. Ellipse; centre (0, 1); major axis $x + 3y - 3 = 0$, length 6; minor axis $3x + y - 1 = 0$, length $2\sqrt{6}$.
5. A pair of straight lines, $x - 2y - 2 = 0$ and $x + y + 1 = 0$.
6. Hyperbola; centre (– 2, 2); transverse axis $x - y + 4 = 0$, length $2\sqrt{2}$; conjugate axis $x + y = 0$, length $2\dfrac{\sqrt{2}}{5}$.
7. Rectangular hyperbola, centre (4, 2); transverse axis $x - 2y = 0$ length $2\sqrt{10}$; asymptotes $3x - y = 10$ and $x + 3y = 10$.
8. Ellipse; centre $\left(\dfrac{3}{5}, \dfrac{1}{5}\right)$; major axis $4x + 3y - 3 = 0$ length 4/5; minor axis $3x - 4y - 1 = 0$, length $\dfrac{4}{15}\sqrt{6}$.
9. Hyperbola; centre (2, – 1); transverse axis $x + 3y + 1 = 0$, length 2; conjugate axis $3x - y = 7$, length 1; asymptotes $x + y = 1$, $x - 7y = 9$.
10. Ellipse; centre $\left(-\dfrac{13}{25}, \dfrac{1}{25}\right)$; major axis $3x + 4y + 2 = 0$, length $2\sqrt{5}$; minor axis $4x - 3y + 3 = 0$, length $\sqrt{10}$.
11. Hyperbola; centre (3, 4); transverse axis $4x + 3y - 24 = 0$, length 2; conjugate axis $3x - 4y - 7 = 0$, length $4\sqrt{6}$; $e = 5$.
12. Hyperbola; centre (1, 0); transverse axis $x - 2y - 1 = 0$, length 4; conjugate axis $2x + y - 2 = 0$, length 6; Asymptotes $8x - y - 8 = 0$ and $4x + 7y - 4 = 0$; foci $\left(1 \pm 2\sqrt{13/5}, -\sqrt{13/5}\right)$.

13. Ellipse, centre $(2, -3)$; major axis $4x + 3y + 1 = 0$, length 6; major axis $3x + 4y + 18 = 0$, length 4; major axis meets the conic in $\left(\frac{1}{5}, -\frac{3}{5}\right)$ and $\left(\frac{19}{5}, -\frac{27}{5}\right)$.

14. Hyperbola; centre $(-1, 2)$; transverse axis $x + 2y = 3$, length $\sqrt{2}$; conjugate axis $y - 2x - 4 = 0$, length $\frac{2}{3}\sqrt{3}$.

15. Hyperbola; centre $(-1, 1)$; transverse axis $x + 7y = 6$, length 2; conjugate axis $7x - y + 8 = 0$, length $\frac{\sqrt{14}}{5}$. The curve cuts the axis of x at $\left(-\frac{31}{12}, 0\right)$ and axis of y at $\left(0, \frac{17}{12}\right)$.

3.14 Tracing of the Parabola

The given equation of second degree i.e.

$$ax^2 + 2hxy + by^2 + 2gx + 2fy + c = 0 \quad \text{...(1)}$$

will represent a parabola, if,

$$h^2 = ab \text{ and } \Delta = 0.$$

Let us put $a = \alpha^2$ and $b = \beta^2$ so that $h = \alpha\beta$.

Then the equation (1) will reduce to the form,

$$(\alpha x + \beta y)^2 = -(2gx + 2fy + c) \quad \text{...(2)}$$

Introducing a constant λ in (2) such that,

$$(\alpha x + \beta y + \lambda)^2 = -2gx - 2fy - c + 2\lambda(ax + \beta y) + \lambda^2$$

$$\Rightarrow \quad (\alpha x + \beta y + \lambda)^2 = 2x(\lambda a - g) + 2y(\lambda\beta - f) + (\lambda^2 - c) \quad \text{...(3)}$$

Now, selecting λ in such a manner that the lines represented by,

$$\alpha x + \beta y + \lambda = 0 \text{ and}$$

$$2x(\lambda\,\alpha x - g) + 2y(\lambda\beta - f) + (\lambda^2 - c) = 0$$

may represent two perpendicular straight lines, which will be taken as the axis and the tangent at the vertex of the parabola, respectively.

Hence, $\left(-\frac{\alpha}{\beta}\right)\left(-\frac{\lambda\alpha - g}{\lambda\beta - f}\right) = -1$

$$\Rightarrow \quad \alpha(\lambda\alpha - g) + \beta(\lambda\beta - f) = 0$$

$$\Rightarrow \quad \lambda = \frac{\alpha g + \beta f}{\alpha^2 + \beta^2} \quad \text{...(4)}$$

In order to reduce the equation of the parabola in standard form i.e., $y^2 = 4ax$, rearrange equation (3) in the form

$$\left(\frac{\alpha x+\beta y+\lambda}{\sqrt{\alpha^2+\beta^2}}\right)^2(\alpha^2+\beta^2)$$

$$= 2\sqrt{\{(\alpha\lambda-g)^2+(\beta\lambda-f)^2\}}\left[\frac{2x(\alpha\lambda-g)+2y(\beta\lambda-f)+(\lambda^2-c)}{2\sqrt{\{(\alpha\lambda-g)^2+(\beta\lambda-f)^2\}}}\right]$$

or
$$\left(\frac{\alpha x+\beta y+\lambda}{\sqrt{\alpha^2+\beta^2}}\right)^2=\left[\frac{2\sqrt{\{(\lambda\alpha-g)^2+(\lambda\beta-f)^2\}}}{\alpha^2+\beta^2}\right]\left[\frac{2x(\lambda\alpha-g)+2y(\lambda\beta-f)+(\lambda^2-c)}{2\sqrt{\{(\lambda\alpha-g)^2+(\lambda\beta-f)^2\}}}\right]$$

Putting, $\dfrac{\alpha x+\beta y+\lambda}{\sqrt{\alpha^2+\beta^2}}=Y$

$$\frac{2x(\lambda\alpha-g)+2y(\lambda\beta-f)+(\lambda^2-c)}{2\sqrt{(\lambda\alpha-g)^2+(\lambda\beta-f)^2}}=X$$

and
$$\frac{2\sqrt{(\lambda\alpha-g)^2+(\lambda\beta-f)^2}}{\alpha^2+\beta^2}=4a$$

The equation of the conic in standard form becomes

$$Y^2 = 4aX.$$

Equation of the axis will be, then

$$Y=0$$

i.e.
$$\frac{\alpha x+\beta y+\lambda}{\sqrt{\alpha^2+\beta^2}}=0$$

$$\Rightarrow \quad \alpha x+\beta y+\lambda=0 \qquad ...(5)$$

and, equation of the tangent at the vertex will be,

$$X=0$$

$$\Rightarrow 2x(\lambda\alpha-g)+2y(\lambda\beta-f)+(\lambda^2-c)=0 \qquad ...(6)$$

where λ is given by (4).

Solving eqns. (5) and (6), we get the vertex of the parabola.

Length of whose latus-rectum is

$$4a = \frac{2\sqrt{(\lambda\alpha - g)^2 + (\lambda\beta - f)^2}}{\alpha^2 + \beta^2}.$$

3.15 The Focus S and Equations of the Directrix and the Latus-rectum

Referred to the reduced equation of the parabola in the form, $Y^2 = 4aX$, we have the following :

Focus : Focus is given by

$$X = a, Y = 0$$

Above will give us two equations in x and y. Solving these equations for x, y, we get the coordinates of the focus, S.

Equation of the directrix referred to new axes will be

$$X = -a$$

Hence, referred to old axes, its equation will be,

$$\frac{\alpha x + \beta y + \lambda}{\sqrt{\alpha^2 + \beta^2}} = -a$$

Equation of latus-rectum : Equation of latus-rectum, referred to new system of axes is given by,

$$X = a$$

Hence, referred to original axes, its equation will be given by

$$\frac{\alpha x + \beta y + \lambda}{\sqrt{\alpha^2 + \beta^2}} = a.$$

Note. In order to trace the shape of the parabola, it is required to find out the points where the parabola meets the coordinate axes. Hence while drawing the figure, let the parabola pass through those points.

ILLUSTRATIVE EXAMPLES

Example 1. Trace the parabola,

$$9x^2 + 24xy + 16y^2 - 2x + 14y + 1 = 0$$

and find the coordinates of its focus and the equation to its directrix.

Solution : The given equation can be re-arranged in the form :

$$(3x + 4y)^2 = 2x - 14y - 1 \qquad ...(i)$$

Introducing a λ in it, we get

$$(3x + 4y + \lambda)^2 = x(6\lambda + 2) + y(8\lambda - 14) + \lambda^2 - 1 \quad ...(ii)$$

Choose λ, such that the straight lines,

$$3x + 4y + \lambda = 0 \text{ and}$$

$$x(6\lambda + 2) + y(8\lambda - 14) + (\lambda^2 - 1) = 0$$

are at right angles.

Hence, $\left(-\frac{3}{4}\right)\left(-\frac{6\lambda + 2}{8\lambda - 14}\right) = -1$

$\Rightarrow \qquad \lambda = 1$

Hence, equation of the conic reduces in the form

$$(3x + 4y + 1)^2 = 8x - 6y$$

Rearranging, we get

$$25\left(\frac{3x + 4y + 1}{\sqrt{3^2 + 4^2}}\right)^2 = \frac{2(4x - 3y)}{\sqrt{4^2 + 3^2}} \times 5$$

$$\Rightarrow \qquad \left(\frac{3x + 4y + 1}{5}\right)^2 = \frac{2}{5}\left(\frac{4x - 3y}{5}\right)$$

Let, $\qquad \frac{3x + 4y + 1}{5} = Y$

and $\qquad \frac{4x - 3y}{5} = X$

The above equation reduces in the form,

$$Y^2 = \frac{2}{5}X$$

which is standard equation of a parabola.

Length of **latus-rectum** i.e. $4a = \frac{2}{5}$

$\therefore \qquad a = \frac{1}{10}.$

Equation of **axis** of the parabola in reference to new system of axes will be

$$Y = 0$$

Hence, referred to old axes it will be

$$\frac{3x + 4y + 1}{5} = 0$$

$\Rightarrow \qquad 3x + 4y + 1 = 0 \qquad ...(iii)$

And, similarly, equation of the tangent at the vertex will be

$$X = 0$$

$\Rightarrow \qquad 4x - 3y = 0 \qquad ...(iv)$

Solving eqns. (iii) and (iv), we get the coordinates of the vertex as $\left(-\frac{3}{25}, -\frac{4}{25}\right)$.

Also, the focus of the parabola will be given by

$$X = a, Y = 0$$

i.e. $\quad \frac{4x - 3y}{5} = \frac{1}{10}$ and $3x + 4y + 1 = 0$

i.e. $\quad 8x - 6y - 1 = 0$ and $3x + 4y + 1 = 0$

Solving these, we get the **focus** at $\left(-\frac{1}{25}, -\frac{11}{50}\right)$

Equation of the directrix is given by,

$$X + a = 0$$

$\Rightarrow \qquad \frac{4x - 3y}{5} + \frac{1}{10} = 0$

$\Rightarrow \qquad 8x - 6y + 1 = 0.$

Tracing : Where the parabola meets the axis of x, $y = 0$. Hence the given equation gives,

$$9x^2 - 2x + 1 = 0$$

which gives imaginary values of x. It shows that parabola does not intersect the axis of x. Similarly, where it meets the axis of y, $x = 0$. Hence, $16y^2 + 14y + 1 = 0$. Solving we get $y = -1$ and -8 approximately.

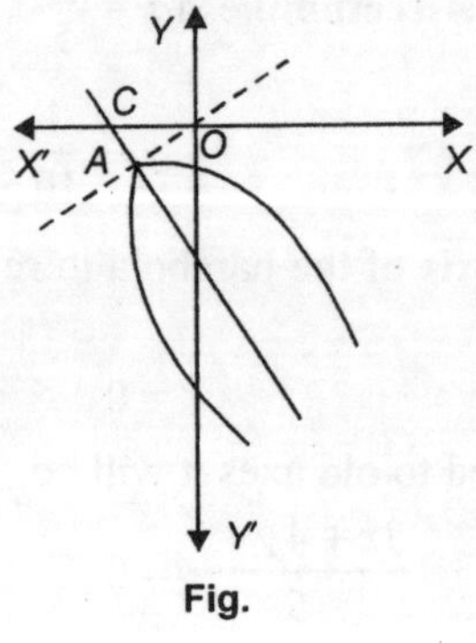

Fig.

Taking the vertex as $A\left(-\frac{3}{25},-\frac{4}{25}\right)$.

The axis passes through A. Let it meet the axis of x in C, then at C, $y = 0$. Hence (iii) gives

$$3x + 1 = 0$$

$$\Rightarrow \qquad x = -\frac{1}{3}.$$

Hence coordinates of C will be $\left(-\frac{1}{3},0\right)$. It also passes through focus $S\left(-\frac{1}{25},-\frac{11}{25}\right)$.

Hence, the shape of the curve (parabola) will be as given in the figure.

Example 2. Trace the conic, $x^2 + 4y^2 - 4xy - 32x + 4y + 16 = 0$.

Solution : Since second degree terms in the given equation,

$$x^2 - 4xy + 4y^2 - 32x + 4y - 16 = 0 \qquad \text{...(i)}$$

of the conic form a perfect square and $\Delta \neq 0$ (calculate it), hence, the conic is a parabola. Rearranging the given equation, we get

$$(x - 2y)^2 - 32x + 4y + 16 = 0 \qquad \text{...(ii)}$$

which, on introducing λ becomes,

$$(x - 2y + \lambda)^2 = (2\lambda + 32)\,x - (4\lambda + 4) + (\lambda^2 - 16) \qquad \text{...(iii)}$$

Choose λ, such that, the straight lines,

$$x - 2y + \lambda = 0 \text{ and}$$

$$(2\lambda + 32)\,x - (4\lambda + 4)\,y + (\lambda^2 - 16) = 0$$

are at right angles. Hence,

$$\left(\frac{1}{2}\right)\left(\frac{2\lambda + 32}{4\lambda + 4}\right) = -1$$

$$\Rightarrow \qquad \lambda = -4$$

Hence, (iii) reduces in the form

$$\left(\frac{x - 2y - 4}{\sqrt{5}}\right)^2 = \frac{12}{\sqrt{5}}\left(\frac{2x + y}{\sqrt{5}}\right)$$

$$\Rightarrow \qquad Y^2 = \frac{12}{\sqrt{5}}X,$$

where $$X = \frac{2x + y}{\sqrt{5}}$$

and $$Y = \frac{x - 2y - 4}{\sqrt{5}}$$

Hence, length of latus-rectum of the parabola i.e. $4a = \dfrac{12}{\sqrt{5}}$

$$\Rightarrow \qquad a = \frac{3}{\sqrt{5}}.$$

Axis of the parabola is

$$Y = 0$$

$$\Rightarrow \qquad x - 2y - 4 = 0 \qquad \text{...(iv)}$$

and tangent at vertex is

$$X = 0$$

$$\Rightarrow \qquad 2x + y = 0 \qquad \text{...(v)}$$

Solving eqns. (iv) and (v), we get the coordinates of the vertex of the parabola at $\left(\frac{4}{5}, -\frac{8}{5}\right)$

Its **focus** is given by

$$X = a, Y = 0$$

$$\Rightarrow \qquad 2x + y = 3, x - 2y = 4.$$

Solving these equations, the coordinates of the focus are (2, – 1).

Directrix

It is given by $\qquad X = -a$

$$\Rightarrow \qquad \frac{2x + y}{\sqrt{5}} + \frac{3}{\sqrt{5}} = 0$$

$$\Rightarrow \qquad 2x + y + 3 = 0.$$

Tracing : Take the vertex $A\left(\frac{4}{5}, -\frac{8}{5}\right)$.

Through A, draw a line AC inclined at an angle $\tan^{-1}\left(\frac{1}{2}\right)$, given by (4), to X-axis. This line is the axis of the parabola. Through A draw another line at right angles to AC. It is the tangent at the vertex.

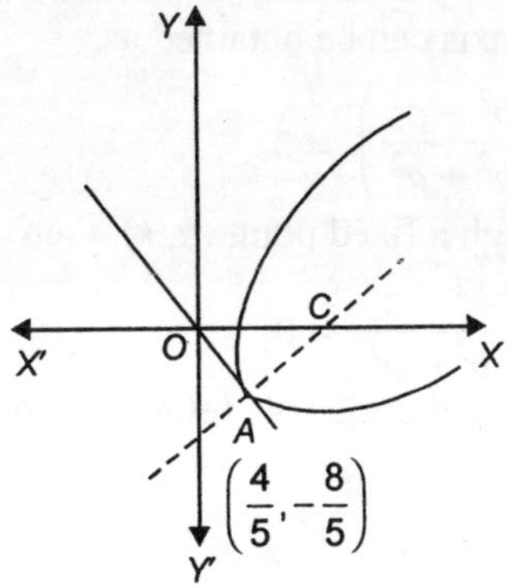

Fig.

We can very easily show that the parabola meets the axis of x at (31.2, 0) and (·8, 0), approximately and it meets the axis of y in imaginary point.

Hence the shape of the conic will be as given in the figure.

EXERCISE 3 (B)

1. Trace the parabola,

$$16x^2 - 24xy + 9y^2 + 77x - 64y + 95 = 0.$$

Also find the coordinates of its focus.

2. Find the axis, the vertex, the latus-rectum and the focus of the parabola,

$$16x^2 - 24xy + 9y^2 - 104x - 172y + 44 = 0.$$

3. Trace the conic $(x + 2y)^2 + 12x - 6y = 0$. Find its focus and the equation of the latus-rectum.

Trace the conics :

4. $4x^2 - 4xy + y^2 - 8x - 6y + 50 = 0.$

5. $9x^2 - 24xy + 16y^2 - 18x - 101y + 19 = 0.$

6. $x^2 + 2xy + y^2 - 2x - 1 = 0.$

7. Obtain the focus, directrix, and axis of the parabola $\sqrt{ax} + \sqrt{by} = 1$. If its axis passes through a fixed point, show that the locus of the focus is a rectangular hyperbola.

[***Hint.*** Equation of the parabola can be simplified in the form,

$$(ax - by)^2 = 2ax + 2by - 1.$$

Equation of the axis can be obtained as,

$$ax - by + \left(\frac{b^2 - a^2}{b^2 + a^2}\right) = 0.$$

If it passes through a fixed point (h, k), then

$$ah - bk + \frac{b^2 - a^2}{b^2 + a^2} = 0 \qquad ...(2)$$

The coordinates of the focus is referred to original axes can be written as $\left(\frac{a}{a^2 + b^2}, \frac{b}{a^2 + b^2}\right)$.

If (x, y) be the focus, then,

$$x = \frac{a}{a^2 + b^2}, y = \frac{b}{a^2 + b^2}$$

Hence, $$x^2 + y^2 = \frac{1}{a^2 + b^2}$$

$\Rightarrow$ $$a^2 + b^2 = \frac{1}{x^2 + y^2}$$

Putting this value in the coordinates of the focus, we get

$$a = \frac{x}{x^2 + y^2}, b = \frac{y}{x^2 + y^2}$$

Eliminating a, b from (α) and above, we get on simplification

$$x^2 - y^2 - hx + ky = 0$$

as locus of the focus. This equation, clearly represents a rectangular hyperbola.]

8. Find the length of latus rectum of the parabola,

$$(a^2 + b^2)(x^2 + y^2) = (bx + ay - ab)^2.$$

9. If t be a variable quantity, then, find the locus of the point (x, y), when, $x = at + bt^2, y = bt + at^2$.

ANSWERS
Exercise 3(B)

1. Axis $4x - 3y + 10 = 0$; tangent at vertex $3x + 4y + 5 = 0$; length of L.R. $\frac{1}{5}$ focus $\left(-\frac{217}{190}, \frac{11}{25}\right)$.

2. $4x - 3y + 2 = 0$; $\left(\frac{1}{5}, \frac{2}{5}\right)$; 8, $3x + 4y = 11$; (1, 2).

3. Axis $x + 2y = 0$, tangent at vertex $2x = y$, latus-rectum $2x - y + \frac{3}{2} = 0$, length $\frac{6}{5}\sqrt{5}$, focus $\left(-\frac{3}{5}, \frac{3}{10}\right)$.
4. Axis $2x + y = 1$, vertex $\left(\frac{3}{5}, \frac{1}{5}\right)$, focus $\left(\frac{4}{5}, \frac{3}{5}\right)$, latus-rectum $4/\sqrt{5}$.
5. Axis $3x - 4y + 7 = 0$, tangent at vertex $4x + 3y + 2 = 0$, latus-rectum 3.
6. Axis $2x + 2y - 1 = 0$, tangent at vertex $4x - 4y + 5$, latus-rectum $\frac{1}{\sqrt{2}}$.

4

System of Conics

4.1 Conic through Five Points

To show that one and only one conic can be drawn through five given points, no three of which are collinear.

The general equation of the conic

$$ax^2 + 2hxy + by^2 + 2gx + 2fy + c = 0$$

contains five arbitrary constants.*

In order to evaluate these constants, we require five independent relations between them. The coordinates of every one of these five points, when substituted in the general equation of the conic lead to five independent relations between the arbitrary constants therein.

Every one of the five relations between the arbitrary constants will be seen to be linear. They will thus lead to only one set of values for these constants and therefore only one conic will be found satisfying the given condition.

4.2 Conic through Four Points and Touching a Straight Line

A conic is fixed if four points through which it passes and one straight line which it touches are known.

The coordinates of the four points lead to four independent relations between the constants. The fifth relation is obtained by the condition of tangency of the straight line. But the fifth will give a quadratic relation. Evaluation of the arbitrary constants from these five relations will therefore lead to more than one set of values for them and consequently more than conic will be found satisfying these conditions.

4.3 To Find the General Equation of the Conic Passing through the Point of Intersection of a Curve and a Straight Line

The general equation of the conic is

$$S \equiv ax^2 + 2hxy + by^2 + 2gx + 2fy + c = 0 \qquad ...(1)$$

---

* Apparently there are six, but we can divide out by any one of them.

and the equation of the straight line is

$$u = lx + my + n = 0 \quad ...(2)$$

Consider the equation

$$ax^2 + 2hxy + by^2 + 2gx + 2fg + c + \lambda (lx + my + n) = 0 \quad ...(3)$$

This equation being of second degree represents a conic. The coordinates of the points which satisfy (1) and (2) also satisfy (3). Therefore the points of intersection of (1) and (2) lie on (3).

Therefore, (3) is the equation of a conic passing through the points of intersection of (1) and (2)

i.e. $$S + \lambda u = 0$$

Example. If $y = 2x$ be the equation of the chord of a circle $x^2 + y^2 = 10x$. Find the equation of the circle of which this chord is a diameter.

Solution : The equation of any circle passing through the ends of the chord intercepted on $y = 2x$ by the circle

$$x^2 + y^2 = 10x$$

is $$x^2 + y^2 - 10x + \lambda (y - 2x) = 0 \quad ...(i)$$

or $$x^2 + y^2 - 2x (5 + \lambda) + \lambda y = 0$$

Coordinates of the centre of this circle is $\left(5 + \lambda, -\frac{\lambda}{2}\right)$.

These coordinates should satisfy the chord $y = 2x$, therefore

$$-\frac{\lambda}{2} = 2 (5 + \lambda)$$

or $$\lambda = -4$$

Substituting this value of λ in (i), we get the required equation of the circle

$$x^2 + y^2 - 2x (5 - 4) - 4y = 0$$

or $$x^2 + y^2 - 2x - 4y = 0.$$

4.4 To Find the Equations to the Conic Sections Passing through the Intersection of a Conic and Two Given Straight Lines

Let the equation of the conic section be

$$S \equiv ax^2 + by^2 + 2hxy + 2gx + 2fy + c = 0 \quad ...(1)$$

and the equations of the straight lines be

$$\mu_1 \equiv l_1 x + m_1 y + n_1 = 0 \quad ...(2)$$

and $$\mu_2 \equiv l_2x + m_2y + n_2 = 0 \quad ...(3)$$

Conic (1) is cut by the line (2) in the points P and Q and by (3) in points R and T respectively.

The equation of a conic passing through their four points of intersection can be written as

$$S + \lambda u_1u_2 = 0 \quad ...(4)$$

For different values of λ, we shall obtain different conics passing through these above points.

4.5 Intersection of Two Conics

To show that two conics in general meet in FOUR points, real or imaginary.

Let the equation of the two conics be

$$S_1 \equiv a_1x^2 + 2h_1xy + b_1y^2 + 2g_1x + 2f_1y + c_1 = 0 \quad ...(1)$$

and $$S_2 \equiv a_2x^2 + 2h_2xy + b_2y^2 + 2g_2x + 2f_2y + c_2 = 0 \quad ...(2)$$

since both the equations are of second degree, therefore on solving one of these equations for y in terms of x and substituting in the other equation, we get an equation of the fourth degree in x giving four values of x, real or imaginary.

If we eliminate y^2 between the equations, we obtain y in terms of x and x^2 which shows that corresponding to one value of x, there will be one value of y.

Similarly by eliminating x from (1) and (2), a biquadratic in y can also be obtained which gives four values of y real or imaginary and their corresponding values of x can be determined.

Hence in general, two conics intersect in four points real or imaginary.

Note. Either all the values of x (or y) are real or an even number of them are imaginary.

4.6 To Find the General Equation of the Conic Passing through the Points of Intersections of Two Conics

$$S_1 \equiv a_1x^2 + 2h_1xy + b_1y^2 + 2g_1x + 2f_1y + c_1 = 0 \quad ...(1)$$

$$S_2 \equiv a_2x^2 + 2h_2xy + b_2y^2 + 2g_2x + 2f_2y + c_2 = 0 \quad ...(2)$$

The equation of the conic passing through the intersection of (1) and (2) is

$$S_1 + \lambda S_2 = 0 \quad ...(3)$$

where λ is an arbitrary constant. Since the equations (1) and (2) are of second degree, equation (3) will also be of second degree and hence it will represent a conic passing through the point of intersection of (1) and (2) for all values of λ.

Example. Find the equation of the two circles passing through the points of intersection of circles $x^2 + y^2 + 2x + 3y = 7$ and $x^2 + y^2 - 6x + 2y = 5$ and also passing through $(2, -3)$.

Solution : The equation of any curve passing through the points of intersection of the given circles is

$$(x^2 + y^2 + 2x + 3y - 7) + \lambda (x^2 + y^2 - 6x + 2y - 5) = 0 \quad ...(i)$$

This circle passes through $(2, -3)$, so we have

$$(4 + 9 + 4 - 9 - 7) + \lambda (4 + 9 - 12 - 6 - 5) = 0$$

or $$\lambda = \frac{1}{10} \quad ...(ii)$$

Substituting the value of λ in (i), we get

$$10 (x^2 + y^2 + 2x + 3y - 7) + (x^2 + y^2 - 6x + 2y - 5) = 0$$

or $11x^2 + 11y^2 + 14x + 32y = 75$.

4.7 Some Special Properties of Circles

Different cases of intersection of two circles

Let $C_1 (h_1, k_1)$ and $C_2 (h_2, k_2)$ be the centres and a_1 and a_2 be the radii of the two circles respectively. The following cases of intersection of two circles arise.

Case I. ***Circles may not cut in real points***

(a) *When the circles are entirely apart from each other.*

Here the distance between their centres will be greater than the sum of the radii

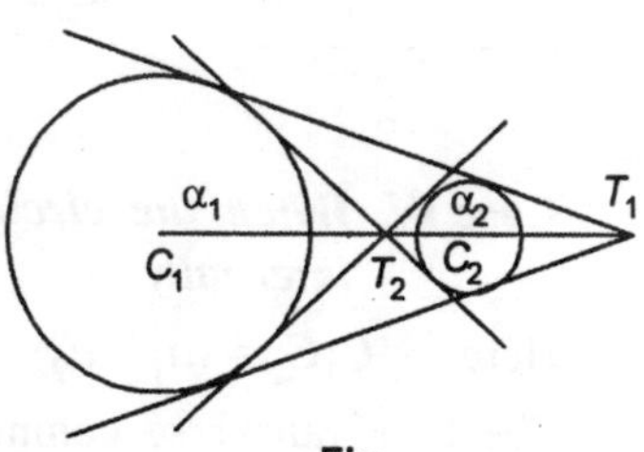

Fig.

$$C_1C_2 > (a_1 + a_2)$$

i.e., $(h_1 - h_2^2) + (k_1 - k_2^2) > (a_1 + a_2)$

Here the tangents meeting the line of centres at T_1 are known as **direct common tangents** and those meeting at T_2 are known as **transverse common tangents.**

Note. T_1 and T_2 divide C_1C_2, the line of centres externally and internally in the ratio of their radii.

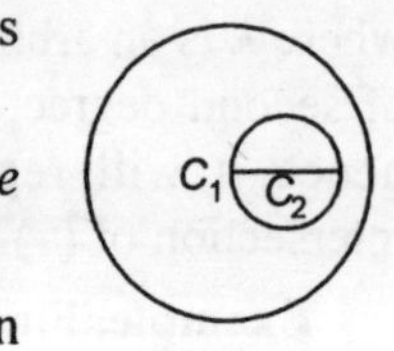

Fig.

(b) *When one circle lies completely within the other.*

Here $C_1C_2^2 < (a_1 - a_2)^2$ and all the four common tangents are imaginary.

Case II. ***When the circles touch each other externally***

Here

$$C_1C_2 = a_1 + a_2$$

The two direct common tangents are real and distinct and the transverse common tangents coincide.

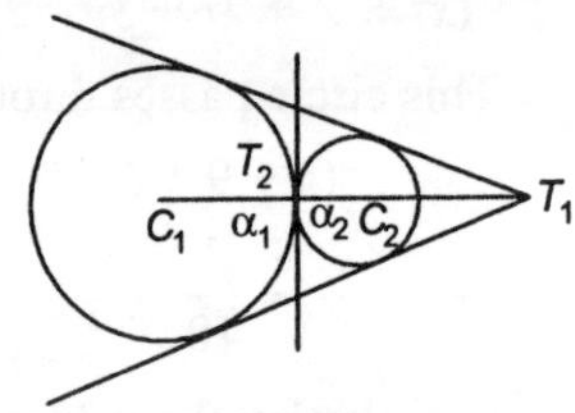

Fig.

Case III. ***When the circles intersect in real points***

Here $\quad C_1C_2 < (a_1 + a_2)$

The two direct common tangents are real and distinct, but the transverse common tangents are imaginary.

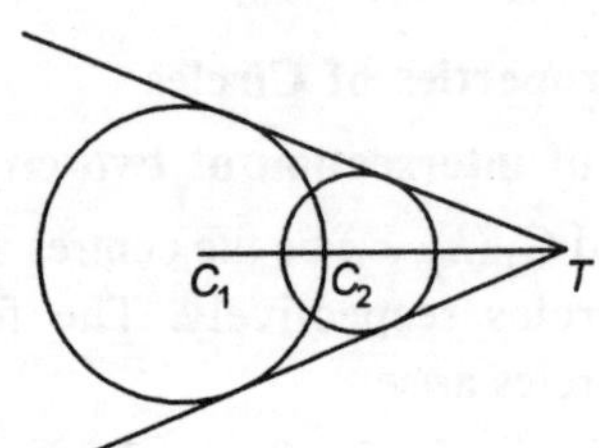

Fig.

Case IV. ***When the circles touch each other internally***

Here $\quad C_1C_2 = (a_1 - a_2)$

The two transverse common tangents coincide and the direct common tangents are imaginary.

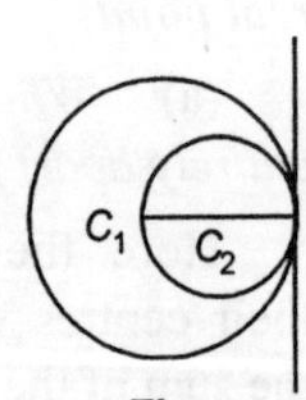

Fig.

Example. Show that the circles

$$x^2 + y^2 + 2x - 8y + 8 = 0$$

and $\quad x^2 + y^2 + 10x - 2y + 22 = 0$

touch one another, find the equation of their common tangents and the point of contact.

Solution : The equation to the circles are

$$x^2 + y^2 + 2x - 8y + 8 = 0 \quad ...(i)$$

and $$x^2 + y^2 + 10x - 2y + 22 = 0 \quad ...(ii)$$

Centre of the circle (i) is (– 1, 4) and radius = 3

Centre of the circle (ii) is (– 5, 1) and radius = 2

Distance between their centres = $\sqrt{(-1+5)^2 + (4-1)^2} = 5$

Sum of their radii = 3 + 2 = 5

Since distance between their centres is equal to sum of the radii, therefore, the circle touch one another externally.

Subtraction gives

$$4x + 3y + 7 = 0 \quad ...(iii)$$

which is the required equation of the common tangent.

For point of contact, solve (iii) with (i) or (ii).

From (iii) $$y = -\frac{4}{3}(x + 2)$$

Substituting the value of y in (1), we obtain the value of x. Hence the point of contact is $\left(-\frac{17}{5}, \frac{1}{5}\right)$.

4.8. Common Chord of Two Circles

Definition : The straight line joining the point of intersection (real or imaginary) of two circles is called the common chord of the two circles.

4.9 To Find the Equation of the Common Chord of Two Circles

$$S_1 \equiv x^2 + y^2 + 2g_1x + 2f_1y + c_1 = 0 \quad ...(1)$$

and $$S_2 \equiv x^2 + y^2 + 2g_2x + 2f_2y + c_2 = 0 \quad ...(2)$$

$S_1 - S_2$ gives

$$(g_1 - g_2) + (f_1 - f_2)\,y + c_1 - c_2 = 0 \quad ...(3)$$

Equation (3) being of the first degree, represents the straight line and satisfied by the same value of x and y which satisfy (1) and (2) simultaneously. Therefore, (3) represents the required equation of the common chord.

Note. Coefficients of x^2 and y^2 must be converted to unity, if not so, by dividing the whole equation by the coefficient of x^2.

4.10 To Find the Length of the Common Chord of the Two Circles

In the adjoining figure

$$AB = 2AM$$
$$= 2\sqrt{AC_1^2 - C_1^2M}$$

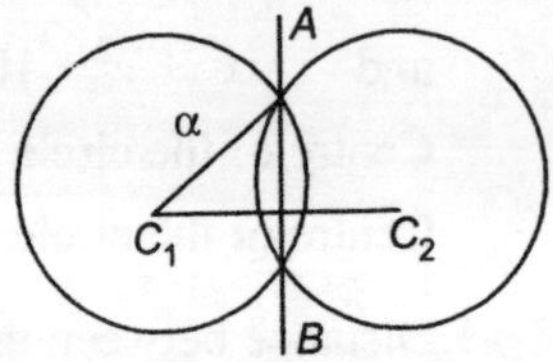

Fig.

where AC_1 is the radius of the circle and C_1M is the length of the perpendicular from centre upon the common chord.

4.11 Equation of Common Chord

When the length of the common chord is zero, i.e. B approaches to A, the two circles touch each other and the common chord becomes the common tangent to the circles at the point of contact.

Example. Find the equation and the length of the chord of the two circles given by the equations

$$x^2 + y^2 + 2x + 2y + 1 = 0 \quad \text{...(i)}$$

and $x^2 + y^2 + 4x + 3y + 2 = 0$ ...(ii)

Solution : The equation of the common chord is

$$2x + y + 1 = 0 \quad \text{...(iii)}$$

Centre of the circle (i) is (– 1, – 1) and its radius is 1.

Therefore, the length of the perpendicular from (– 1, – 1) to $2x + y + 1 = 0$ is

$$p = \pm\left(\frac{-2-1+1}{\sqrt{4+1}}\right) = \frac{2}{\sqrt{5}}$$

Therefore, length of the common chord $= 2\sqrt{(\text{radius})^2 - p^2}$

$$= 2\sqrt{1 - \frac{4}{5}} = \frac{2}{\sqrt{5}}.$$

EXERCISE 4 (A)

1. Show that the circle on the chord $x \cos \alpha + y \sin \alpha = p$ of the circle $x^2 + y^2 = a^2$ as diameter is
$$x^2 + y^2 - a^2 - 2p\,(x \cos \alpha + y \sin \alpha - p) = 0.$$
2. If $y = mx$ be the equation of a chord of the circle $x^2 + y^2 - 2ax = 0$, prove that the equation of the circle of which this chord is the diameter is
$$(1 + m^2)(x^2 + y^2) - 2a\,(x + my) = 0.$$

3. Prove that the equation of the circle on the straight line joining the points of intersection of $ax^2 + 2hxy + by^2 = 0$ and $lx + my = 1$ as diameter is

$$(x^2 + y^2)(am^2 - 2hlm + bl^2) + 2x(hm - bl) + 2y(hl - am) + (a + b) = 0.$$

4. Determine the positions of the following pairs of circles with respect to each other
   (a) $x^2 + y^2 = 12$ and $x^2 + y^2 - 5x + 3y - 2 = 0$
   (b) $x^2 + y^2 - 4x - 10y + 4 = 0$ and $x^2 + y^2 - 6x - 12y - 55 = 0$.
5. Find the equations to the four common tangents to the circles $x^2 + y^2 + 2x - 8y + 13 = 0$ and $x^2 + y^2 - 12x - 14y + 76 = 0$.
6. Prove that the two circles $x^2 + y^2 + 2ax + c = 0$ and $x^2 + y^2 + 2by + c = 0$

   touch each other if $\dfrac{1}{a^2} + \dfrac{1}{b^2} = \dfrac{1}{c}$.
7. Find the equation to the common tangent of the circles :
   (a) $x^2 + y^2 = 2$; $x^2 + y^2 - 6x - 6y + 10 = 0$.
   (b) $x^2 + y^2 = 1$; $(x - 1)^2 + (y - 3)^2 = 4$.
   (c) $x^2 + y^2 - 2x - 6y + 9 = 0$; $x^2 + y^2 + 6x - 2y + 1 = 0$.
   (d) $x^2 + y^2 - 2x = 0$; $x^2 + y^2 + 6x - 6y + 2 = 0$.
   (e) $x^2 + y^2 - 4x + 6y + 8 = 0$; $x^2 + y^2 - 10x - 6y + 14 = 0$.
8. Find the equation of the circle whose radius is 3 and which touches the circle $x^2 + y^2 - 4x - 6y - 12 = 0$ internally at the point $(-1, -1)$.
9. Find the equation and the length of the common chord of the two circles :
   (a) $x^2 + y^2 + 2x + 3y + 1 = 0$; $x^2 + y^2 + 4x + 3y + 2 = 0$.
   (b) $x^2 + y^2 - 4x + 8y - 18 = 0$; $2x^2 + 2y^2 + 7x - 5y + 2 = 0$.
   (c) $(x - a)^2 + (y - b)^2 = c^2$; $(x - b)^2 + (y - a)^2 = c^2$.
10. Find the equation of the circle drawn on the common chord of the following pair of circles as diameter :
    (a) $x^2 + y^2 - 6x - 4y = 12$; $x^2 + y^2 - 2x - 6y = 15$
    (b) $x^2 + y^2 + 2x + 3y + 1 = 0$
    $x^2 + y^2 + 4x + 3y + 2 = 0$
    (c) $(x - a)^2 + y^2 = a^2$; $x^2 + (y - b)^2 = b^2$.

ANSWERS

Exercise 4(A)

4. (a) Intersect at real points.

(b) The first circle lies completely within the other circle.

5. $y - \dfrac{26}{5} = \dfrac{21 \pm 5\sqrt{33}}{24}\left(x - \dfrac{9}{5}\right); y + 2 = \dfrac{21 \pm \sqrt{57}}{48}(x + 15)$

7. (a) $x + y = 2$.

(b) $x + 1 = 0$; $4x - 3y - 5 = 0$; $y - 1 = 0$ and $3x + 4y = 5$.

(c) $x = 0$; $3x + 4y = 10$; $y = 4$ and $3y = 4x$.

(d) $4x - 3y + 1 = 0$. (e) $x + 2y - 1 = 0$.

8. $\left(x - \dfrac{4}{5}\right)^2 + \left(y - \dfrac{7}{5}\right)^2 = 3^2$.

9. (a) $2x + 1 = 0$; $2\sqrt{2}$.

(b) $5x - 21y + 38 = 0; 2\sqrt{1102/333}$.

(c) $x - y = 0; \sqrt{4c^2 - 2(a - b)^2}$.

10. (a) $2x^2 + 2y^2 - 8x - 10y - 27 = 0$.

(b) $2x^2 + 2y^2 + 2x + 6y + 1 = 0$.

(c) $(a^2 + b^2)(x^2 + y^2) = 2ab(bx + ay)$.

4.12 Angle of Intersection of Two Curves

Definition : If the curves C_1 and C_2 intersect at P, and PT_1 and PT_2 be the tangents to the two curves at P, then the angle between the tangents i.e., T_1PT_2 is called the angle of intersection of the two curves at the point of intersection.

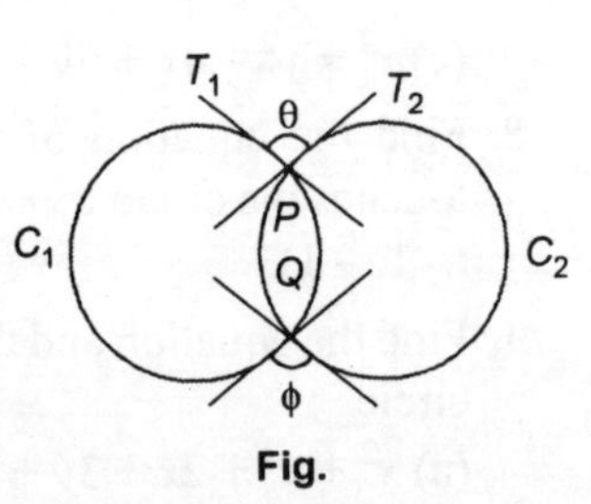

Fig.

4.13 Orthogonal Curves

Definition : When the two curves intersect at an angle of 90°, they are said to intersect **orthogonally,** and the curve themselves are called orthogonal curves.

4.14 Condition of Orthogonality

To find the condition that the two circles

$$x^2 + y^2 + 2g_1x + 2f_1y + c_1 = 0$$

and $x^2 + y^2 + 2g_2x + 2f_2y + c_2 = 0$ may be orthogonal.

Consider, the centres C_1 and C_2 of the circles are $(-g_1, -f_1)$ and $(-g_2, -f_2)$ and their corresponding radii are

$$\sqrt{f_1^2 + g_1^2 - c_1} \text{ and } \sqrt{f_2^2 + g_2^2 - c_2}$$

Let P be one of the point of intersection and PT_1 and PT_2 are tangents drawn to the circles. Since the circles intersect at an angle of 90° i.e.

$$\angle C_1PC_2 = 90°$$

Therefore, the tangents pass through the centres of the circle. Join C_1C_2. Now in the right angled triangle C_1PC_2, we have

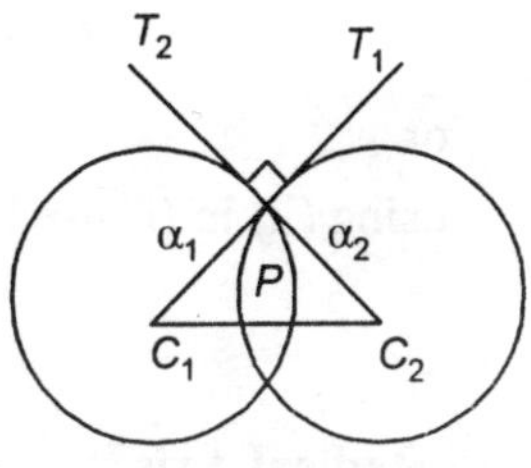

Fig.

$$C_1C_2^2 = C_1P^2 + C_2P^2$$

$$(g_1 - g_2)^2 + (f_1 - f_2)^2 = g_1^2 + f_1^2 - c_1 + g_2^2 + f_2^2 - c_2$$

or $\quad 2(g_1g_2 + f_1f_2) = c_1 + c_2$

which is the required condition.

4.15 Angle of Intersection of Two Circles

Since the tangents at P are perpendicular to their radii C_1P and C_2P, therefore, the angle of intersection of the two circles is equal to the angle between their radii at the point of intersection. But from trigonometry, we have

$$\cos C_1PC_2 = \frac{C_1P^2 + C_2P^2 - C_1C_2^2}{2 \cdot C_1P \cdot C_2P}$$

Example. Prove that the two circles which pass through $(0, a)$, and $(0, -a)$ and touch the straight line $y = mx + c$ will cut orthogonally, if

$$c^2 = a^2(2 + m^2)$$

Solution : Since the circles pass through $(0, a)$ and $(0, -a)$, therefore, the centre will lie on x-axis. Let the coordinates of the centres be $(h, 0)$. The radius of the circle is equal to the distance of $(h, 0)$ from either of the given point $(0, a)$ and $(0, -a)$

i.e., $\sqrt{h^2 + a^2}$

If the line $y = mx + c$ is the tangent to such circles, we have radius equal to the length of the perpendicular from centre upon the line $y = mx + c$

$$\sqrt{h^2 + a^2} = \frac{mh + c}{\sqrt{1 + m^2}}$$

or $$(h^2 + a^2)(1 + m^2) = (mh + c)^2$$

or $h^2 - 2mhc + a^2(1 + m^2) - c^2 = 0$

This is a quadratic in h and will give two values of h say h_1 and h_2

$$h_1 h_2 = a^2(1 + m^2) - c^2 \qquad ...(i)$$

But the circles cut orthogonally, therefore, sum of the squares of their radii is equal to the square of the distance between the centres.

$$(h_1 - h_2)^2 = h_1^2 + a^2 + h_2^2 + a^2$$

or $$h_1 h_2 = -a^2 \qquad ...(ii)$$

using (ii) in (i), we have

$$-a^2 = a^2(1 + m^2) - c^2$$

or $$c^2 = a^2(2 + m^2).$$

4.16 Radical Axis

Definition : The locus of a point, which moves so that the length of the tangents drawn from it to two circles are equal is called the radical axis.

4.17 To find the equation of the Radical Axis of the Circles

$$x^2 + y^2 + 2g_1x + 2f_1y + c_1 = 0$$

and $$x^2 + y^2 + 2g_2x + 2f_2y + c_2 = 0$$

Let the moving point be (h, k) through which the tangents are drawn to the circles. By hypothesis, the squares of their lengths will be equal

$$h^2 + k^2 + 2g_1h + 2f_1k + c_1 = h^2 + k^2 + 2g_2h + 2f_2k + c_2$$

or $2h(g_1 - g_2) + 2k(f_1 - f_2) + (c_1 - c_2) = 0$

Generalising the coordinates, we get the required locus of (h, k).

$$2x(g_1 - g_2) + 2y(f_1 - f_2) + c_1 - c_2 = 0$$

which represents a straight line.

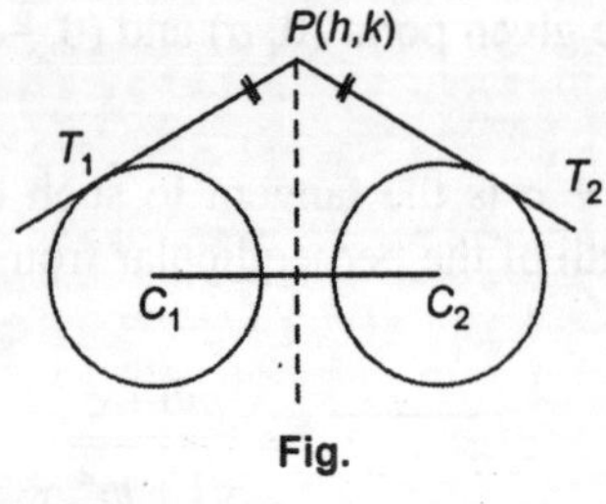

Fig.

4.18 Radical Axis and Common Chord of the Two Circles

If the equations to the two circles be expressed as

$S_1 = 0$ and $S_2 = 0$ [the coeff. of x^2 = coeff. of $y^2 = 1$]

We have the equation of the radical axis as

$$S_1 - S_2 = 0$$

Since $S_1 - S_2 = 0$ represents a curve passing through the points of intersections of $S_1 = 0$ and $S_2 = 0$.

Thus the radical axis and the common chord of two circles are same.

Ex. 1. Prove that the radical axis of the two circles and the line of centres are perpendicular to each other.

Ex. 2. Show that the radical axis of two circles bisects each of their common tangent.

4.19 Circles Intersecting Two Given Circles Orthogonally

Let the equation of the circles be $S_1 = 0$ and $S_2 = 0$ ...(1)

The equation of the radical axis is $S_1 - S_2 = 0$ ...(2)

Let (x_2, y_2) be any point on the radical axis,

then we have

$$2\,(g_1 - g_2)\,x_1 + 2\,(f_1 - f_2)\,y_1 + c_1 - c_2 = 0 \qquad ...(3)$$

Now the length of the tangent from (x_1, y_2) to first circle

$$= \sqrt{x_1^2 + y_1^2 + 2g_1x_1 + 2f_1y_1 + c_1} \qquad ...(4)$$

The equation of the circle with centre (x_1, y_1) and radius equal to the length of the tangent i.e. (4), is

$$(x - x_1)^2 + (y - y_1)^2 = x_1^2 + y_1^2 + 2g_1x_1 + 2f_1y_1 + c_1$$

i.e., $x^2 + y^2 - 2xx_1 - 2yy_1 - (2g_1x_1 + 2f_1y_1 + c_1) = 0$ ...(5)

which intersect the given circles orthogonally, since

$$2g_1\,(-x_1) + 2f_1\,(-y_1) = c_1 - (2g_1x_1 + 2f_1y_1 + c_1)$$

Thus any circle having its centre at a point on the radical axis of the given circles and with radius equal to the length of tangent from this point to either of these circles, cuts them orthogonally.

4.20 To Prove that the Radical Axis of Three Circles, Taken Two at a Time, Meet in a Point Provided the Centres, of the Circles Not Collinear.

Let the equation of the three circles be

$$S_1 = x^2 + y^2 + 2g_1x + 2f_1y + c_1 = 0 \qquad ...(1)$$

$$S_2 = x^2 + y^2 + 2g_2 x + 2f_2 y + c_2 = 0 \quad ...(2)$$

and $S_3 = x^2 + y^2 + 2g_3 x + 2f_3 y + c_3 = 0 \quad ...(3)$

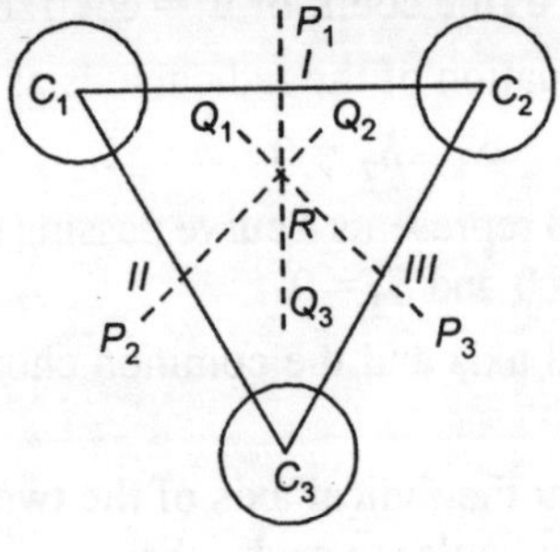

Fig.

The radical axes of the circles taken in pairs are

$$S_1 - S_2 \equiv 2(g_1 - g_2)x + 2(f_1 - f_2)y + c_1 - c_2 = 0 \quad ...(4)$$

$$S_2 - S_3 \equiv 2(g_2 - g_3)x + 2(f_2 - f_3)y + c_2 - c_3 = 0 \quad ...(5)$$

and $S_3 - S_1 \equiv 2(g_3 - g_1)x + 2(f_3 - f_1)y + c_3 - c_1 = 0 \quad ...(6)$

Adding the left hand sides, we find that their sum identically vanishes.

Hence the three lines represented by these three equations are concurrent.

4.21 Radical Centre

Definition : The point of concurrence of the radical axes of three circles taken in pairs is called the radical centre of the three circles.

or

The point from where the length of the tangents to the three circles are equal is called the radical centre.

4.22 To Prove the Circle having Centre at the Radical Centre and Radius Equal to the Length of the Tangent from it to any Circle will Cut all the Three Circles Orthogonally

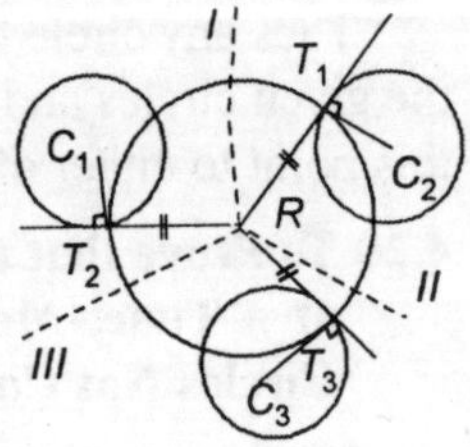

Fig.

Let R be the radical centre of the three circles. Draw RT_1, RT_2 and RT_3 tangents from R to the three circles. By definition $RT_1 = RT_2 = RT_3$. Now draw a circle with R as centre and RT_1 as radius. RT_1 is a tangent to the circle C_2 and C_2T_1 is the radius of the circle.

Clearly RT_1 and C_2T_1 are the tangents at the point of intersection of the circle C_2 and the circle with R as centre inclined at an angle of 90° and thus cut orthogonally.

Similarly the circle with R as centre also cuts the other circles orthogonally.

ILLUSTRATIVE EXAMPLES

Example 1. If two circles intersect a third orthogonally, prove that their radical axis pass through the centre of the third circle.

Solution : Let the equations of the circle be

$S_r \equiv x^2 + y^2 + 2g_r x + 2f_r y + c_r = 0, (r = 1, 2, 3)$

S_1 and S_2 intersect orthogonally

$$\therefore \quad 2g_1g_3 + 2f_1f_3 = c_1 + c_3 \qquad ...(1)$$

S_2 and S_3 intersect orthogonally,

$$\therefore \quad 2g_2g_3 + 2f_2f_3 = c_2 + c_3 \qquad ...(2)$$

Subtracting eqn. (2) from (1), we obtain

$$2(g_1 - g_2)g_3 + 2(f_1 - f_2)f_3 = c_1 - c_2 \qquad ...(3)$$

Now the radical axis of S_1 and S_2 is

$$2(g_1 - g_2)x + 2(f_1 - f_2)y + c_1 - c_2 = 0 \qquad ...(4)$$

Condition (3) shows that (4) should pass through $(-g_3, -f_3)$ which is the centre of the circle S_2.

Example 2. Find the equation of the circle orthogonal to the two circle $x^2 + y^2 - 6x + 8 = 0$ and $x^2 + y^2 - 2x - 2y - 7 = 0$ and passing through the origin.

Solution : The equation of the two circles are

$$x^2 + y^2 - 6x + 8 = 0 \qquad ...(i)$$

$$\text{and } x^2 + y^2 - 2x - 2y - 7 = 0 \qquad ...(ii)$$

Let the equation of the required circle be

$$x^2 + y^2 + 2gx + 2fy + c = 0 \qquad ...(iii)$$

Since it passes through (0, 0), therefore,

$$c = 0$$

$\therefore$ The equation (iii) becomes

$$x^2 + y^2 + 2gx + 2fy = 0 \qquad ...(iv)$$

$\because$ (iv) cuts (i) orthogonally

$$\therefore \qquad -6g = 8 \Rightarrow g = -\frac{4}{3} \qquad ...(v)$$

Similarly (iv) cuts (ii) orthogonally

$$\therefore \qquad -2g - 2f = -7$$

$$\Rightarrow \qquad 2f = \frac{29}{8} \text{ [using (v)]} \qquad ...(vi)$$

Hence the required equation of the circle is

$$3(x^2 + y^2) - 8x + 29y = 0.$$

Example 3. Find the radical centre and the equation to a circle which intersects these circles orthogonally

$$x^2 + y^2 + 2x + 17y + 4 = 0 \qquad ...(i)$$

$$x^2 + y^2 + 7x + 6y + 11 = 0 \qquad ...(ii)$$

$$x^2 + y^2 - x + 22y + 3 = 0 \qquad ...(iii)$$

Solution : Radical axis of (i) and (ii) is

$$-5x + 11y - 7 = 0 \qquad ...(iv)$$

Again radical axis of (ii) and (iii) is

$$8x - 16y + 8 = 0 \qquad ...(v)$$

Solving eqns. (iv) and (v), we have $x = 3$ and $y = 2$

The length of the tangent from (3, 2) to any of the circles is equal to

$$\sqrt{(9 + 4 + 6 + 34 + 4)} = \sqrt{(57)}$$

Now the equation of the circle with centre (3, 2) and radius $\sqrt{(57)}$ is

$$(x - 3)^2 + (y - 2)^2 = 57$$

$$\text{or} \qquad x^2 + y^2 - 6x - 4y = 44.$$

Example 4. A circle cuts the rectangular hyperbola $xy = 1$ in the points (x_r, y_t); $r = 1, 2, 3, 4$; prove that

$$x_1x_2x_3x_4 = y_1y_2y_3y_4 = 1$$

Solution : Let the equation of the circle of radius a be

$$(x - g)^2 + (y - f)^2 = a^2 \qquad ...(i)$$

and the rectangular hyperbola is $xy = 1$ ...(ii)

Eliminating y in eqns. (i) and (ii), we obtain

$$(x - g)^2 + \left(\frac{1}{x} - f\right)^2 = a^2$$

or $$x^2(x-g)^2 + (1-fx)^2 = a^2x^2$$

or $$x^4 - 2xg^3 + x^2(g^2+f^2-a^2) - 2fx + 1 = 0$$

This is a fourth degree equation in x giving four values of x, say x_1, x_2, x_3 and x_4. From the theory of equations, the product of the roots is given by

$$x_1x_2x_3x_4 = 1$$

To every value of x, the corresponding value of y can be calculated from eqn. (ii) and thus

$$y_1 y_2 y_3 y_4 = \frac{1}{x_1x_2x_3x_4} = 1.$$

EXERCISE 4 (B)

1. Find the equation to the radical axis of the circles :

(a) $x^2 + y^2 + ax + by + c = 0$; $x^2 + y^2 + bx + ay + c = 0$.

(b) $2x^2 + 2y^2 + 14x - 18y + 15 = 0$; $4x^2 + 4y^2 - 3x - y + 5 = 0$.

2. Find the equation of the circles which pass through the common points of the following pair circles :

(a) $x^2 + y^2 + 2x + 3y - 7 = 0$ and $x^2 + y^2 + 3x - 2y - 1 = 0$ through the point (1, 2).

(b) $x^2 + y^2 + 4x - 2y = 0$ and $x^2 + y^2 - 2x + 4y = 0$ and touching the line $x + y = 1$.

(c) $x^2 + y^2 + 4x - 6x - 12 = 0$ and $x^2 + y^2 - 5x + 7y = 19$ and having its centre on $x + y = 0$.

3. If the two conics whose equations referred to rectangular axes are

$$ax^2 + 2hxy + by^2 + 2gx + 2fy + c = 0$$

and $$a'x^2 + 2h'xy + b'y^2 + 2g'x + 2f'y - c' = 0$$

intersect in four concyclic points, prove that

$$\frac{a-b}{h} = \frac{a'-b'}{h'}$$

4. Find the condition that the circle $x^2 + y^2 + 2g_1x + 2f_1y + c_1 = 0$ shall bisect the circumference of the circle

$$x^2 + y^2 + 2g_2x + 2f_2y + c_2 = 0.$$

5. Find the equation of the circle which pass through $(2a, 0)$ and whose radical axis in relation to the circle $x^2 + y^2 = a^2$ is $x = \frac{1}{2}a$.

6. Prove that the square of the tangent that can be drawn from any point of one circle to another circle is equal to twice the product of the perpendicular distance of the point from the radical axis of the two circles and the distance between the centres.

7. Find the radical centre of the following set of circles :

(a) $x^2 + y^2 - 16x + 60 = 0$; $x^2 + y^2 - 12x + 27 = 0$;
$x^2 + y^2 - 12y + 84 = 0$.

(b) $x^2 + y^2 + 4x + 7 = 0$; $2(x^2 + y^2) + 3x + 5y + 9 = 0$;
$x^2 + y^2 + y = 0$

(c) $x^2 + y^2 - 3x - 6y + 14 = 0$; $x^2 + y^2 - x - 4y + 8 = 0$;
$x^2 + y^2 + 2x - 6y + 9 = 0$

(d) $3x^2 + 3y^2 + 4x - 6y - 1 = 0$; $2x^2 + 2y^2 - 3x - 2y - 4 = 0$;
$2x^2 + 2y^2 - x + y - 1 = 0$.

8. Find the point from which the tangents to the following circles are equal in length $x^2 + y^2 - 16x + 16 = 0$; $3x^2 + 3y^2 - 36x + 8 = 0$ and $x^2 + y^2 - 16x - 12y + 8 = 0$. Also find the length of this tangent.

9. Prove that the following pair of circles cut orthogonally :

(a) $x^2 + y^2 - 2px - 2qy = 2pq$;
$x^2 + y^2 + 2qx + 2py = 2pq$

(b) $x^2 + y^2 - 2x - 6y + 11 = 0$;
$x^2 + y^2 - 10x - 4y + 21 = 0$

(c) $x^2 + y^2 - 6x - 8y + 24 = 0$;
$x^2 + y^2 - 8x - 6y + 24 = 0$.

10. Find the equation to the circle orthogonal to the two circles :

(a) $x^2 + y^2 - 9x + 14 = 0$; $x^2 + y^2 + 15x + 14 = 0$
and passing through the point (2, 5).

(b) $x^2 + y^2 - 6y + 1 = 0$; $x^2 + y^2 - 4y + 1 = 0$
and touch the line $3x + 4y + 5 = 0$.

(c) $x^2 + y^2 - 4x - 6y + 11 = 0$; $x^2 + y^2 - 10x - 4y + 21 = 0$
and has $2x + 3y = 7$ as a diameter.

11. Find the equation to the circles, cutting orthogonally each of the following circles :

(a) $x^2 + y^2 - 2x + 3y - 7 = 0$; $x^2 + y^2 + 5x - 5y + 9 = 0$;
$x^2 + y^2 + 7x - 9y + 29 = 0$

(b) $x^2 + y^2 - x + y = 0$; $2x^2 + 2y^2 + 5x - 3y - 2 = 0$;
$x^2 + y^2 + 4x - 2y - 2 = 0$

(c) $x^2 + y^2 = a^2$; $(x - c)^2 + y^2 = a^2$; $x^2 + (y - b)^2 = a^2$.

PARABOLA

12. A circle and a parabola intersect in four points; show that the algebraic sum of the ordinates of the four points is zero.
Show also that the line joining one pair of these four points and the line joining the other pair are equally inclined to the axis.

HYPERBOLA

13. Prove that all conics which pass through the intersection of two rectangular hyperbola are themselves rectangular hyperbolas.

14. If a circle cuts a rectangular hyperbola $xy = c^2$ in A, B, C, D and the parameters of these four points be t_1, t_2, t_3, t_4 respectively; prove that

(i) $t_1, t_2, t_3, t_4 = 1$

(ii) the centre of the mean position of the four points bisects the distance between the centres of the two curves.

15. A rectangular hyperbola whose centre is C is cut by any circle of radius r in the four points P, Q, R, S; prove that

$$CP^2 + CQ^2 + CR^2 + CS^2 = 4r^2$$

16. A circle intersects a hyperbola in four points; prove that the product of the distances of the four points of intersection from one asymptote is equal to the product of their distances from the other.

17. A rectangular hyperbola $xy = c^2$ is cut by a circle passing through its centre O in the points A, B, C, D whose parameters are t_1, t_2, t_3, t_4; show that

$$(t_1 + t_2)(t_3 + t_4) = -(t_1, t_2, t_3, t_4)$$

and deduce that the product of the perpendiculars from O on AB and CD is c^2.

ANSWERS

Exercise 4(B)

1. (a) $x - y = 0$; (b) $31x - 35y + 15 = 0$

2. (a) $x^2 + y^2 + 4x - 7y + 5 = 0$

(b) $x^2 + y^2 + (1 - \sqrt{3})x + (1 + \sqrt{3})y = 0$

$x^2 + y^2 + (\sqrt{3} - 1)x + \sqrt{3}(\sqrt{3} - 1)y = 0.$

(c) $2(x^2 + y^2) - x + y = 31 = 0$

4. $2g_2(g_1 - g_2) + 2f_2(f_2 - f_1) = c_1 - c_2$

5. $x^2 + y^2 - 2ax = 0$

7. (a) $\left(\frac{33}{4}, 13\right)$ (b) $(-2, -1)$

(c) $(1, 2)$ (d) $\left(-\frac{16}{21}, \frac{-31}{63}\right)$

8. $\left(\frac{10}{3}, -\frac{2}{3}\right); \frac{2}{3}\sqrt{58}.$

10. (a) $x^2 + y^2 - 3y - 14 = 0;$ (b) $4x^2 + 4y^2 - 15x = 4;$

(c) $x^2 + y^2 - 4x - 2y + 3 = 0.$

11. (a) $x^2 + y^2 - 16x - 18y - 4 = 0;$ (b) $x^2 + y^2 - 2(x + y) = 0;$

(c) $x^2 + y^2 - cx - by + a^2 = 0.$

4.23 Confocal Conics

Definition : Two Conics are said to be confocal if their both foci are common.

Let, $$\frac{x^2}{a^2} + \frac{y^2}{b^2} = 1 \quad ...(1)$$

be the equation of a given conic. Here it is an ellipse, hence its foci will lie at $\left(\pm\sqrt{a^2 - b^2}, 0\right)$.

Let the equation of the conic, confocal to (1) be

$$\frac{x^2}{A^2} + \frac{y^2}{B^2} = 1 \quad ...(2)$$

Then its foci will lie at $\left(\pm\sqrt{A^2 - B^2}, 0\right)$.

Since both the conics are confocal i.e. have the same foci, hence

$$a^2 - b^2 = A^2 - B^2$$

$$\Rightarrow \quad A^2 - a^2 = B^2 - b^2 = \lambda \text{ (say)}$$

$$\Rightarrow \quad A^2 = a^2 + \lambda \text{ and } B^2 = b^2 + \lambda$$

Hence (2) gives the equation of the confocal conic to (1) for every value of λ, as

$$\frac{x^2}{a^2 + \lambda} + \frac{y^2}{b^2 + \lambda} = 1 \quad ...(3)$$

Thus (3) represents a system of conics, confocal to the ellipse (1) i.e. for different values of λ we get different members of (3) all having their foci at the foci of (1).

λ is called the **parameter** of the system.

The different members of the system given by (3) are shown as in the figure given below :

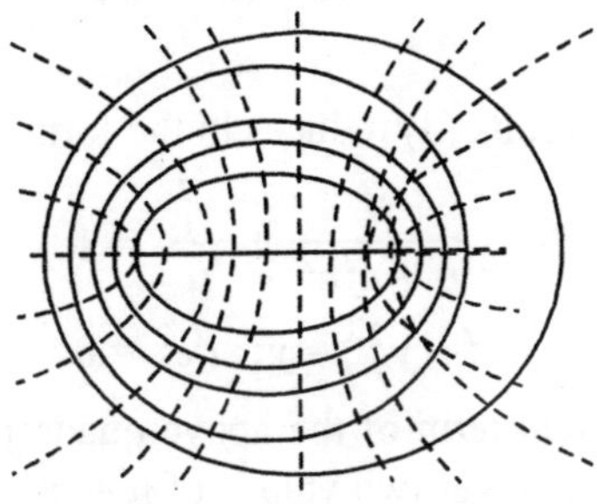

Fig.

4.24 Confocals through a Given Point

Through a given point in the plane of a given conic, there can be two conics passing through the point and confocal to the given conic. One of these will be an ellipse and the other one will be a hyperbola :

Let the given conic be,

$$\frac{x^2}{a^2} + \frac{y^2}{b^2} = 1 \qquad ...(1)$$

and $A(x_1, y_1)$ be the given point. Also, let the equation of the conic confocal to (1) be,

$$\frac{x^2}{a^2+\lambda} + \frac{y^2}{b^2+\lambda} = 1 \qquad ...(2)$$

If this passes through $A(x_1, y_1)$, then

$$\frac{x_1^2}{a^2+\lambda} + \frac{y_1^2}{b^2+\lambda} = 1 \qquad ...(3)$$

Simplifying it, we get

$$x_1^2(b^2+\lambda) + y_1^2(a^2+\lambda) = (a^2+\lambda)(b^2+\lambda)$$

$$\Rightarrow b^2x_1^2 + a^2y_1^2 + \lambda(x_1^2 + y_1^2) = a^2b^2 + \lambda(a^2+b^2) + \lambda^2$$

$$\Rightarrow \lambda^2 + (a^2+b^2-x_1^2-y_1^2)\lambda + (a^2b^2 - b^2x_1^2 - a^2y_1^2) = 0 \qquad ...(4)$$

which is quadratic in λ, hence gives its two values. Corresponding to each value of λ, (2) will give a conic confocal to the given conic. Which shows that through a given point, there will be two confocal conics to the conic (1).

Again, let $\qquad b^2 + \lambda = \mu$

Hence, $\qquad a^2 + \lambda = a^2 + \mu - b^2$

$$= \mu + (a^2 - b^2)$$

$$= \mu + a^2e^2 \qquad \Big| \because b^2 = a^2(1 - e^2)$$

For, the above results (3) reduces in the form

$$\frac{x_1^2}{\mu + a^2e^2} + \frac{y_1^2}{\mu} = 1$$

$$\Rightarrow \quad \mu^2 + \mu(a^2e^2 - x_1^2 - y_1^2) - y_1^2 a^2e^2 = 0 \qquad \text{...(5)}$$

Since, the constant term of the above quadratic equation (5) is negative, hence we will get two values of $\mu = b^2 + \lambda$, of which one will be positive and the other will be negative (by theory of quadratic equation). Similarly we get both the values of $a^2 + \lambda$ positive.

Hence, corresponding to these values of $a^2 + \lambda$ and $b^2 + \lambda$, we will get one confocal as an ellipse and the other one as a hyperbola.

4.25 Confocals Cut at Right Angles

Let the equations of the two confocals be

$$\frac{x^2}{a^2 + \lambda_1} + \frac{y^2}{b^2 + \lambda_1} = 1 \qquad \text{...(1)}$$

and $$\frac{x^2}{a^2 + \lambda_2} + \frac{y^2}{b^2 + \lambda_2} = 1 \qquad \text{...(2)}$$

Let these two curves meet in the point (x_1, y_1), then, we get

$$\frac{x_1^2}{a^2 + \lambda_1} + \frac{y_1^2}{b^2 + \lambda_1} = 1 \qquad \text{...(3)}$$

and $$\frac{x_1^2}{a^2 + \lambda_2} + \frac{y_1^2}{b^2 + \lambda_2} = 1 \qquad \text{...(4)}$$

Subtracting eqn. (4) from (3), we get

$$x_1^2\left(\frac{1}{a^2 + \lambda_1} - \frac{1}{a^2 + \lambda_2}\right) + y_1^2\left(\frac{1}{b^2 + \lambda_1} - \frac{1}{b^2 + \lambda_2}\right) = 0$$

$$\Rightarrow \quad \frac{x_1^2}{(a^2+\lambda_1)(a^2+\lambda_2)} + \frac{y_1^2}{(b^2+\lambda_1)(b^2+\lambda_2)} = 0$$

$$\Rightarrow \quad \frac{x_1^2 (b^2+\lambda_1)(b^2+\lambda_2)}{y_1^2 (a^2+\lambda_1)(a^2+\lambda_2)} = -1 \qquad [\because \lambda_1 \neq \lambda_2] \qquad ...(5)$$

Also, equations of the tangents of (1) and (2) at their common point of intersection (x_1, y_1) will be

$$\frac{x_1 x}{a^2+\lambda_1} + \frac{y_1 y}{b^2+\lambda_1} = 1 \qquad ...(6)$$

and $$\frac{x_1 x}{a^2+\lambda_2} + \frac{y_1 y}{b^2+\lambda_2} = 1 \qquad ...(7)$$

Their gradients are,

$$-\frac{x_1 (b^2+\lambda_1)}{y_1 (a^2+\lambda_1)} \text{ and } \frac{-x_1 (b^2+\lambda_2)}{y_1 (a^2+\lambda_2)}$$

The product of the gradients of the tangents (6) and (7)

$$= -\frac{x_1 (b^2+\lambda_1)}{y_1 (a^2+\lambda_1)} \times -\frac{x_1 (b^2+\lambda_2)}{y_1 (a^2+\lambda_2)}$$

$$= \frac{x_1^2 (b^2+\lambda_1)(b^2+\lambda_2)}{y_1^2 (a^2+\lambda_1)(a^2+\lambda_2)}$$

$$= -1 \qquad \text{[by (5)]}$$

Hence, they include a right angle between them.

Therefore, the two confocals (1) and (2), cut at right angles.

4.26 Only One Member of the System of Confocals will Touch a Given Line

Let the equation of the system of confocals be,

$$\frac{x^2}{a^2+\lambda} + \frac{y^2}{b^2+\lambda} = 1 \qquad ...(1)$$

and $$lx + my + n = 0 \qquad ...(2)$$

be the equation of the given straight line.

Let (x_1, y_1) be the point where (1) touches (2).

Then, (2) will be identical to the equation of the tangent of (1) at (x_1, y_1), which is given by

$$\frac{xx_1}{a^2+\lambda} + \frac{yy_1}{b^2+\lambda} - 1 = 0 \qquad ...(3)$$

Applying the condition of (2) and (3) being identical, we get

$$\frac{\frac{x_1}{a^2+\lambda}}{l} = \frac{\frac{y_1}{b^2+\lambda}}{m} = \frac{-1}{n}$$

$$\therefore \qquad x_1 = \frac{-l(a^2+\lambda)}{n}, \; y_1 = -\frac{m(b^2+\lambda)}{n} \qquad ...(4)$$

Since, (x_1, y_1) lies on (2) also, we have

$$lx_1 + my_1 + n = 0$$

$$\Rightarrow \qquad \frac{l^2(a^2+\lambda)}{n} - \frac{m^2(b^2+\lambda)}{n} + n = 0$$

$$\Rightarrow \qquad (a^2+\lambda)\, l^2 + (b^2+\lambda)\, m^2 = n^2$$

which is an equation of first degree in λ, hence gives its one value only.

Hence only one member of the confocals touches a given line.

ILLUSTRATIVE EXAMPLES

Example 1. If the confocals through (x_1, y_1) to the ellipse

$$\frac{x^2}{a^2} + \frac{y^2}{b^2} = 1$$

are, $\dfrac{x^2}{a^2+\lambda_1} + \dfrac{y^2}{b^2+\lambda_1} = 1$ and $\dfrac{x_1^2}{a^2+\lambda_2^2} + \dfrac{y_1^2}{b^2+\lambda^2} = 1$, then show that

(a) $\dfrac{x_1^2}{a^2} + \dfrac{y_1^2}{b^2} - 1 = -\lambda_1\lambda_2/a^2b^2$;

(b) $x_1^2 + y_1^2 - a^2 - b^2 = \lambda_1\lambda_2$.

Solution : Referred to equation (4) of 4.24, we get

$$\lambda_2 + (a^2 + b^2 - x_1^2 - y_1^2)\,\lambda + (a^2b^2 - b^2x_1^2 - a^2y_1^2) = 0 \qquad ...(i)$$

which is quadratic in λ, hence gives its two values, λ_1 and λ_2, say, hence we get two confocals, of the given ellipse, as

$$\frac{x^2}{a^2+\lambda_r} + \frac{y^2}{b^2+\lambda_r} = 1, \; r = 1, 2$$

Hence, applying the theory of quadratic equations, we get from (i)

$$\lambda_1\lambda_2 = \frac{a^2b^2 - b^2 + x_1^2 - a^2y_1^2}{1}$$

$$\Rightarrow \quad -\frac{\lambda_1\lambda_2}{a^2b^2} = -1 + \frac{x_1^2}{a^2} + \frac{y_1^2}{b^2}$$

i.e. $$\frac{x_1^2}{a^2} + \frac{y_1^2}{b^2} - 1 = \frac{-\lambda_1\lambda_2}{a^2b^2}$$

which proves (a).

Again, the sum of roots of (i) are given by,

$$\lambda_1 + \lambda_2 = \frac{-(a^2 + b^2 - x_1^2 - y_1^2)}{1}$$

$$\Rightarrow \quad x_1^2 + y_1^2 - a^2 - b^2 = \lambda_1 + \lambda_2$$

which proves (b).

Example 2. Find the conics confocal with $x^2 + 2y^2 = 2$, which pass through the point (1, 1).

Solution : Equation of the given conic is $x^2 + 2y^2 = 2$

$$\Rightarrow \quad \frac{x^2}{2} + \frac{y^2}{1} = 1 \qquad \text{...(i)}$$

Let, the equation of the system of confocal conics of (i) be

$$\frac{x^2}{2+\lambda} + \frac{y^2}{1+\lambda} = 1 \qquad \text{...(ii)}$$

Since, it pass through the given point (1, 1)

Hence, $$\frac{1^2}{2+\lambda} + \frac{1^2}{1+\lambda} = 1$$

$$\Rightarrow \quad 1 + \lambda + 2 + \lambda = (2 + \lambda)(1 + \lambda)$$

$$\Rightarrow \quad 3 + 2\lambda = 2 + 3\lambda + \lambda^2$$

$$\Rightarrow \quad \lambda^2 + \lambda - 1 = 0$$

which on solving gives,

$$\lambda = \frac{-1 \pm \sqrt{1+4}}{2}$$

$$= \frac{-1 \pm \sqrt{5}}{2}$$

$$\therefore \quad \lambda_1 = \frac{-1+\sqrt{5}}{2}, \lambda_2 = \frac{-1-\sqrt{5}}{2}$$

and the given point be (h, k), then the chord of contact of (h, k) w.r.t. (i) will be given by,

$$\frac{hx}{a^2+\lambda}+\frac{ky}{b^2+\lambda}=1 \qquad \text{...(ii)}$$

From eqns. (i) and (ii), we get

$$a^2+\lambda=\frac{x(hy-kx)}{y-k},\ b^2+\lambda=\frac{hy-kx}{h-x}$$

Subtracting, we get

$$a^2-b^2=(hy-kx)\left(\frac{x}{y-k}-\frac{y}{h-x}\right)$$

which on simplifying gives

$$(a^2-b^2)(y-k)(h-x)=(hy-kx)(hx-x^2-y^2+ky)$$

which is a cubic passing through the given point (h, k) and the foci $\left(\pm\sqrt{a^2-b^2},0\right)$.

Again, if the given point lies on the major axis, then $k = 0$ and the above equation reduces to the form,

$$(a^2-b^2)(h-x)=h(hx-x^2-y^2)$$

$$\Rightarrow h(x^2+y^2)-(a^2-b^2+h^2)x+h(a^2-b^2)=0$$

Evidently, it represents a circle.

Example 6. Prove that the two conics $ax^2+2hxy+by^2=1$ and $a'x^2+2h'xy+b'y^2=1$ can be placed so as to be confocal if

$$\frac{(a-b)^2+4h^2}{(ab-h^2)^2}=\frac{(a'-b')^2+4h'^2}{(a'b'-h'^2)^2}.$$

Solution : Since we know that length of the semi-axis of the conic $ax^2+2hxy+by^2=1$ are given by $\left(a-\frac{1}{r^2}\right)\left(b-\frac{1}{r^2}\right)=h^2$

$$\Rightarrow \quad (ab-h^2)r^4-(a+b)r^2+1=0 \qquad \text{...(i)}$$

which is quadratic in r^2. Hence its roots are, if r_1^2 and r_2^2, then

$$r_1^2+r_2^2=\frac{a+b}{ab-h^2}$$

and

$$r_1^2 r_2^2=\frac{1}{ab-h^2}$$

Also, the distance between the foci of the conic

$$= 2\sqrt{\left(r_1^2 - r_2^2\right)}$$

$$= 2\sqrt{\left(r_1^2 - r_2^2\right)^2 - 4r_1^2 - r_2^2}$$

$$= 2\sqrt{\left(\frac{a+b}{ab-h^2}\right)^2 - \left(\frac{4}{ab-h^2}\right)}$$

$$= 2\sqrt{\frac{(a-b)^2 + 4h^2}{(ab-h^2)^2}}$$

Similarly, the distance between the foci of the other conic will also be equal to $2\sqrt{\dfrac{(a'-b')^2 + 4h'^2}{(a'b'-h'^2)^2}}$

Since the two conics can be placed as confocals if they have coincident pair of foci.

Consequently the distances between their foci will be equal.

Hence,

$$2\sqrt{\frac{(a-b)^2 + 4h^2}{(ab-h^2)^2}} = 2\sqrt{\frac{(a'-b')^2 + 4h'^2}{(a'b'-h'^2)^2}}$$

$$\Rightarrow \qquad \frac{(a-b)^2 + 4h^2}{(ab-h^2)^2} = \frac{(a'-b')^2 + 4h'^2}{(a'b'-h'^2)^2}$$

EXERCISE 4 (C)

1. Prove that the difference of squares of the perpendiculars drawn from the centre upon parallel tangents to two given confocal conics is constant.
2. Prove that the locus of the points lying on a system of confocal ellipses, which have the same eccentric angle α, is a confocal hyperbola whose asymptotes are inclined at an angle 2α.
3. Show that only one of a given system of confocals can have a given straight line as a normal.
4. Prove that the locus of the feet of the normals, drawn from the fixed point (h, k) to each of the series of confocals given by the equation,

$$\frac{x^2}{a^2+\lambda}+\frac{y^2}{b^2+\lambda}=1 \text{ is the cubic curve,}$$

$$\frac{x}{y-k}+\frac{y}{x-h}=\frac{a^2-b^2}{hy-kx}.$$

5. Tangents are drawn to the parabola $y^2 = 4x\sqrt{a^2-b^2}$ and on each is taken the point at which it touches one of the confocals

$$\frac{x^2}{a^2+\lambda}+\frac{y^2}{b^2+\lambda}=1;$$

Prove that locus of such points is a straight line.

6. The locus of the pole of any tangent to a confocal with respect to any circle, whose centre is one of the foci, is obtained and found to be a circle. Prove that, if the circle corresponding to each confocal be taken, they are all co-axal.

7. A series of parallel tangents is drawn to a system of confocals. Prove that the locus of the points of contact is a rectangular hyperbola.

8. Show that the ends of the equal conjugate diameters of a series of confocal ellipses are on a confocal rectangular hyperbola.

9. Show that the locus of the point from which two perpendicular tangents are drawn to two conics one to each of the two given confocals is a circle.